Emigrating Beyond Earth
Human Adaptation and Space Colonization

Cameron M. Smith and Evan T. Davies

Emigrating Beyond Earth

Human Adaptation and Space Colonization

 Springer

Published in association with
Praxis Publishing
Chichester, UK

Dr Cameron M. Smith
Department of Anthropology
Portland State University
Portland
Oregon 97210
U.S.A.

Dr Evan T. Davies
The Explorers Club
46 East 70th Street
New York
New York 10021
U.S.A.

SPRINGER–PRAXIS BOOKS IN POPULAR SCIENCE
SUBJECT *ADVISORY EDITOR*: Stephen Webb, B.Sc., Ph.D., M.Inst.P., C.Phys.

ISBN 978-1-4614-1164-2   ISBN 978-1-4614-1165-9 (eBook)
DOI 10.1007/978-1-4614-1165-9
Springer New York Heidelberg Dordrecht London

Library of Congress Control Number: 2012932259

Cover design: Jim Wilkie
Project copy editor: Stephen Webb
Typesetting: BookEns, Royston, Herts., UK

Printed on acid-free paper

Springer is part of Springer Science+Business Media (www.springer.com)

Contents

Figures

Tables

Acknowledgements and Dedications

We would like to thank the following persons for their comments and guidance, for supply of images, or simply for supporting the concept of this book. Dr. Dava Newman, Massachusetts Institute of Technology Man-Vehicle Laboratory, Dr. Andrew Knoll, Harvard University Department of Palaeontology, Dr. Stephen Webb, University of Portsmouth, Clive Horwood, the Publisher of Praxis Publishing, recipient of the Sir Arthur C. Clarke Award for Achievement in Space Media 2011, Mr. Michael Rubino and Dr. Christopher R. Scotese of the Palaeomap Project.

Cameron M. Smith I dedicate the energies expended in researching and writing this book to the heretical astronomers of Renaissance Europe, who dared to break with the ancient tradition of human centrality in the Universe, and to young people worldwide. I would like to thank my parents for giving me the opportunity to study worldwide, and my professors, fellow students, family, friends and colleagues, for their stimulating discussions. I also thank Ms. Angela Perri for hosting me for several months while I was researching this book at the libraries of Durham University.

Evan T. Davies I dedicate this work to my parents, Mr. Harry Clayton Davies Jr., and Mrs. Margaret Tyler Davies, and again to every good teacher, role model and mentor I have ever had: Mr. Harry Clayton Davies Sr., Dr. Melvin Francis Ames, Mrs. Jane Porzio, Ms. Arvilla Nolan, Mr. Ronnie Cavalieri, Mr. Bill Irwin, MAJ William Maguire USMC, Mr. Charles Mellon, Ms. Elizabeth Frangides-Foster, Mr. Charles Davis, Mr. Harold Birch, Mr. James Jones, Mrs. Ginny Gong, Mrs. Miriam Weisman, Mrs. Muriel Landsberg, Dr. George Williams, Mr. Ernest Meyer, Mrs. Marjorie El-Khadi, Mrs. Marian Cheris, Mrs. Judy Ferris, Mr. Alfred Polukuski, Mr. Louis Troisi, Mr. Blaine Bocarde, Dr. Marina Roseman, Prof. Robert T. Farrell, Dr. W.T.W. Morgan, Prof. Alan Bilsborough, Dr. John Herman Loret, Prof. Sharon Traweek, Prof. Steven A. Tyler and CAPT Stephen P. Hannifin, USN. Each of you chose to spend your lives planting seeds; never knowing what might grow as you moved along. I hope in some small way, you find a reflection of yourselves herein. Also dedicated to the memory of Florence Mary Richardson, who helped me pass many a summer day of my childhood building models of early air and spacecraft, and spending hours with me talking about the classic science fiction novels from the 1940's, 50's and 60's that I discovered in our attic when my family moved to Port Washington; to the memory of Professor Carl Sagan, who beyond inspiring a generation fortunate enough to have learned from him, encouraged his students at Cornell to think critically, to always maintain a healthy skepticism, and keep an eye towards the day when we might leave the cradle and venture from our home world, and to the memory of my childhood friend Joji "George" Mochizuki; who left this Earth far too soon.

"... promises to keep."

Finally, the authors wish to dedicate their effort to every person who chooses to read this book. We hope that our species survives itself; we hope that our children live to see the dawn break on another world, we hope that we ultimately realize our fullest potential as humans, and that one day, our descendants look back from afar on the Earth with gladness, knowing that we made the right choices in our time.

About the Authors

Cameron M. Smith, Ph.D., teaches human evolution and prehistory at the Department of Anthropology at Portland State University in Oregon. His professional training began as a student of Harvard University's early human archaeology field school at the Leakey research station in northern Kenya. After a year at the University of London's Institute of Archaeology, Dr. Smith earned a Joint Honors BA in Anthropology and Archaeology at Durham University before completing graduate degrees in the US and Canada. His courses emphasize adaptation and evolution as structuring factors of human prehistory. Dr. Smith has been widely published in both scientific journals and popular science magazines, including the *American Journal of Physical Anthropology*, *Structure and Dynamics,* the *Journal of Field Archaeology*, *Scientific American MIND*, *Scientific American*, *Evolution: Education and Outreach*, *Archaeology*, *Hang Gliding and Paragliding* and *Spaceflight*. He has written about evolution in books and magazines including *The Top Ten Myths About Evolution* (Prometheus, 2006) endorsed by the National Center for Science Education and the American Library Association, and *The Fact of Evolution* (Prometheus, 2011), endorsed by *Science Daily* and recently picked for the *Scientific American Book Club*. Away from his office, Dr. Smith is an active Scuba diver and paraglider pilot. He is a fellow of both the Royal Geographical Society and the Explorers Club of New York.

Evan T. Davies, Ph.D., began his academic training in archaeology at Cornell University and has conducted fieldwork throughout the United States, Europe, sub-Saharan Africa and the South Pacific. He completed his graduate studies in cultural anthropology at Rice University, where he focused on land use among traditional hunter-gatherer societies in Central Africa. He serves as a defense attaché, African area expert and imagery scientist in the United States Navy. In 2007 while deployed on a combat tour with the multi-national forces in Iraq, Dr. Davies became involved with efforts to preserve Iraqi antiquities and archaeological sites. His experiences in Iraq led him to pursue medical studies and in the coming years he intends to work in wilderness and expedition medicine as well as continue anthropological research into indigenous pharmaceuticals and healing practices. Dr. Davies is a fellow of both the Royal Geographical Society and the Explorers Club of New York and his popular science writing has appeared in Wiley publications as well as *Spaceflight* and *Archaeology* magazines. He has held a lifelong interest in space exploration.

Preface

Discussions of the idea of human space colonization often quickly diverge in one of two ways. Either people agree that it is an interesting and worthwhile endeavor, or they disagree, with the most common argument against it being that humanity has other pressing problems on its hands at the moment, and that space colonization would be callous and immoral because it would consume effort and resources that could be used to address those problems. This is one reason why Gerard K. O'Neill's 1976 book on space colonization, *The High Frontier*, contains large sections dedicated to showing that space colonization would materially benefit people on Earth. We do not feel that such a justification for human space colonization is necessary; the benefit, as we aim to show in this book, will be the preservation of our species itself, by the ancient methods of adaptation and evolution. While we of course believe that human problems on Earth must be addressed, our justification of human space colonization is not that it will immediately or materially help the people of Earth, but that it will provide a greater return: that it will be an insurance policy for our civilization and our species. Considering the dangers to humanity – non-human, such as asteroid impact, or human, such as warfare and the collapse that archaeologists have documented in every civilization of the ancient world – colonizing space would be the act of a mature civilization, and perhaps the only intelligent way to ensure that what humanity has built over the past few millennia will not be lost. Of course, civilization has built horrors as well as libraries, and it can also always be argued that colonizing space would only move human problems off of the Earth. We prefer to be optimistic, however, and we have good reason to believe that this will not be the case.

For many reasons, the first substantial off-Earth colonies will probably be on Mars, and the first colonists to arrive there will look on a world where not a single bullet has been fired, where not a single bomb has ever fallen from the sky. We think early colonists will be very aware of this. They might well prohibit such weapons entirely. Of course, we will still be human, and prone to the same fears and emotions that have governed many of our less-moral choices and actions in the past. But in a pristine world, we think, people will be determined not to repeat old madnesses, and they may be extremely intolerant of even the seeds of division, violence, and waste. In ancient Iceland, murderers and other criminals were banished to the wilderness, where life was nearly impossible. Similar rules may be enforced by early off-Earthers. And if things go badly in off-Earth colonies, for example on Mars, there would remain the solution that has been used by various human groups on Earth for millennia: social fission. In many cultures, it has been customary to relieve tensions by splitting up, an option almost impossible on Earth today. To paraphrase the science fiction writer Robert

Heinlein, *When things get so crowded that you need an ID, it is time to move.* The great thing about space travel is that it has given us somewhere to move. For a long time after the first colonists begin to settle the Red Planet, Mars will offer large landscapes for expansion; places to go if things turn badly.

And, of course, Mars is no end; one of our points in this book will be that human space colonization cannot be thought of as an end, like landing on the Moon nearly 50 years ago. Rather, what we write about in this book is a beginning. What we as anthropologists have learned from the study of human evolution is that as humanity has evolved, it has continually expanded outward, finding new places to live. Evolutionary adaptation to such a wide range of environments as inhabited by our species is the force that has shaped both our biological and cultural evolution. Just as humans on Earth have explored and settled in new environments throughout prehistory, eventually, people of Mars will want to move farther on; in fact, considering what we will describe in this book, that seems both natural and inevitable, while challenging it seems unnatural. We feel that other places in the Solar System, and perhaps some day even the galaxy, and even other galaxies, will all be explored, and some settled. Humanity, by this method of continual expansion, will have aligned itself with the nature of the Universe, which is change. There is no final utopia, because conditions change, and humanity itself changes. If there is a constant it is change and the evolution itself that adapts to that change. But for humanity to engage with this reality, we have to begin somewhere, and we argue that we should begin now. There will never be a best time to begin; it will always be argued that we have more pressing immediate concerns. But remaining focused on those immediate concerns could, in the end, cost us everything. In our daily lives, we invest heavily to protect our future, by buying insurance and minding our doings. Colonizing space will be nothing less than an insurance policy for our species and civilization. It is worth the both the cost and effort.

In addition to arguing that human space colonization is necessary, we argue that it will not be a technocratic endeavor, focused on rockets and robots, but on human beings. Space colonies, after all, will be for people, communities, and cultures. In this way, they will be a continuation of humanity's many adaptations over the last few million years. That the environments we choose to colonize in space will not be much like Earth environments is somewhat immaterial. This is because establishing colonies – and, over time, new branches of humanity – off of Earth will be *adaptation*, the fitting of a life form to a new environment. While human evolution has significant differences from the evolution of many other species in the past few billion years of Earth life, it is still evolution. For this reason, we also argue in this book that the study of evolutionary adaptation will be central to making a success of human space colonization, and near the end of the book we apply the lessons of adaptation in sketching out an adaptive framework for human space colonization.

This book is our attempt to shed new light – derived from understanding evolution and adaptation – on the ages-old dream of human space colonization. We feel that the best way to make it a success will be to use the adaptive lessons of billions of years of Earth life to humanity's next adaptation: the *extraterrestrial* adaptation, extra- referring to outside of or beyond, and terrestrial referring to the planet Earth.

The most recent book to treat our subject from an anthropological perspective was *Interstellar Migration and the Human Experience*, a volume of papers, some written by anthropologists and others with interests in evolution, published in 1985 by anthropologist Ben Finney and Eric M. Jones. That book tackled serious evolutionary issues associated with human space colonization, and remains a good introduction to them, but some of its elements can now be revised with more recent understanding. For example, it was published before the concrete evidence for planets beyond our own Solar System (which are being discovered monthly, now), and several decades of genomic research have vastly improved our understanding of even basic evolutionary principles, and on the macro scale we have new bodies of theory to help understand evolution as well. We also know that the 'dark matter' theory proposed in the volume has been verified, and there is a significant shift today – perhaps unimaginable in the mid-1980s – away from a vision of government-driven human space colonization, and towards private space colonization. Our worries and concerns today also differ from those of the 1980s, at least in part; today climate change is a major concern, having edged out nuclear annihilation, though we have new worries about nuclear terrorism. And while biology and computing were both largely the tools of science in the past, today they are rapidly becoming the tools of a burgeoning biotechnology industry and the 'gaming' entertainment industry, respectively, with numerous and specific attendant effects. In all of these ways, and others, we live in a different world than that of 1985.

Still, many of our suggestions have been offered in the past; in the 1920s, the Russian scientist Konstantin Tsiolkovski was publishing freely about humans living off of the Earth, in space. But in this book we present new arguments for a collective course of action that could ensure the survival of humanity, and, most importantly, we sketch out an evolutionary and adaptive philosophy for space colonization that is informed by the lessons of anthropology and evolution. In part, our book is an attempt to update what was paved out by the authors and editors of *Interstellar Migration and the Human Experience*. But we also write for a new audience, asking them to believe in a course that many will call impossible, and to support the real efforts of small companies and independent researchers who are preparing for the next major adaptive episode of human evolution, which we call the Extraterrestrial Adaptation.

Most space colonization literature has been written by astronomers, engineers and others in the 'hard sciences', and much that they have written has been somewhat technical. But it is equally important to add a human dimension to the discussion of space colonization because space colonization will ultimately be about humanity. Technology must be a tool, not an end.

In this book we are introducing a philosophy of humans-in-space that emphasizes long-term adaptation and evolution, in great contrast to the necessarily short-term, exploratory character of human-in-space activities of the past half century. We comment on occasion on technologies, but our focus is on what we can learn from our own evolutionary background that can be applied in making a success of human space colonization. Overall, our goal is to help shape a philosophy of space colonization rather than to—at this point—define any specific courses of action, such as population sizes for off-Earth colonies, or their cultural and biological composition. Rather, we are suggesting a new character to the concept of humans in space as subtle, but important, as the difference between, say, crossing an ocean by sailboat rather than with a motorized boat, or between climbing a mountain with a small and nimble team as opposed to climbing it with sherpas, supplemental oxygen and so on. The differences are important.

To structure the introduction of humanity to the space colonization literature, we employ the field of anthropology, the evolutionarily-based scientific study of the human species, both biologically and culturally.

We begin by introducing anthropology, showing how it is structured as a discipline, how it generates knowledge, and why it will be significant to the successful human colonization of space. We then discuss arguments for and against space colonization, arguing that it is important, and that we should begin now, and we conclude by sketching out an adaptive framework for human space colonization. Overall, we wish to establish the foundations of a human-centered, evolutionary paradigm for space colonization. Our book is organized in three main sections.

In Part I, *The Context and Uniqueness of Human Evolution and Adaptation,* we place humanity in the larger context of evolution and show why humanity is well-suited for adaptation to space.

We begin with Chapter 1, *The Extraterrestrial Adaptation: Humanity, Evolution and Migration Into Space,* in which we introduce the field of anthropology, describe why and how it will be significant to successful space colonization, and begin to build a framework for thinking about space colonization in terms of evolutionary adaptation.

In Chapter 2, *Stardust: the Origins of Life, Evolution and Adaptation* we summarize the evolutionary history of the genus *Homo,* noting that it is characterized by adaptability.

In Chapter 3, *The Adaptive Suite of the Genus Homo: Symbolism, Language and Niche Construction,* we outline how humanity has adapted to Earth environments throughout prehistory, not because of our biological constitution, but in fact despite it, using the tremendous power of our symbolic, language-using minds to actively alter and construct ecological niches for our own habitation. We also describe the evolution of modern civilization, showing exactly what it is and dispelling the illusion that it is a pinnacle or end to human evolution, and that – as has happened in the past – it could collapse.

In Part II, *Arguments For and Against Human Space Colonization,* we review the arguments for and against human space colonization, concluding that it is not simply an option, but necessary for the survival of our species, or, at least, the survival of civilization, which we feel is worthwhile.

In Chapter 4, *A Choice of Catastrophes: Arguments for Human Space Colonization,* we argue that a number of threats to humanity make it important to begin colonizing space now as an insurance policy for the human species.

In Chapter 5, *Common Objections to Human Space Colonization* we consider the most common objections to space colonization – including that it is too costly, a technocratic stunt, an elitist or escapist endeavor – and recast space colonization as a responsible investment for the future of our species.

In Part III, *Human Adaptation to Space: Lessons from the Past and Shaping the Future,* we reveal an adaptive framework for planning and carrying out human space colonization – informed by evolution at large and the human colonization of difficult environments to date – and discuss some possible future homes for humanity.

We begin with Chapter 6, *Starpaths: Adaptation to Oceania* in which we use an astounding example of human and adaptation in the ancient world to remind readers that human colonization of essentially alien regions is not a new and outlandish endeavor – or one focused on technology – but a natural continuation of millions of years of human adaptation and evolution.

In Chapter 7, *Building an Adaptive Framework for Human Space Colonization,* we outline an evolutionarily-informed, adaptive framework for human space colonization, based on all we have learned about evolutionary adaptation in the last 3+ billion years of Earth life, including human biological and cultural evolution discussed throughout this book.

We conclude with Chapter 8, *Distant Lands Unknown: Informed Speculation on the Human Future in Space,* in which we allow ourselves some informed speculation and offer a vision of how humanity might successfully begin to colonize our Solar System, and how this process would begin to change our species. We examine some of the latest concepts in extra-solar-system travel and interstellar propulsion, ponder the latest evidence of the multitude of worlds that lie beyond our own Solar System, and imagine what, and maybe even who, we might find there.

At the end of each chapter we include notes that often provide interesting details regarding points in the text, and/or references to technical reports, academic journal articles and other material we used to research this book.

Part I

The Context and Uniqueness of Human Evolution and Adaptation

1   The Extraterrestrial Adaptation: Humanity, Evolution, and Migration into Space

"... it is a long way from sailing canoes to interstellar arks. But ever since our ancestors started using tools to survive and eventually flourish in new environments, the pattern of evolution by cultural as well as biological adaptation has been underway. Although the prospect of traveling and living in space might seem 'unnatural' to many, it would represent a logical extension of the technological path our ancestors have been following ..."

Ben R. Finney and Eric M. Jones[1]

This is a book about humans migrating into space, why we believe such migration is important, and how the lessons of anthropology – the study of the human species – will be important to the success of human survival beyond Earth. We have three essential premises.

Premise 1: Human space migration is not optional, but necessary to avoid the extinction of our species.

This premise argues that human space migration must not be considered an option, but mandatory, at least in the long run. We will see in this book that most life that has ever evolved on Earth – up to 99% by some calculations – has become extinct. Numerous threats to humanity's long-term biological survival can be divided into two main classes: extraterrestrial and terrestrial threats.[2]

Extraterrestrial threats include Earth impact by extraterrestrial bodies, such as the comets or bolides that impacted Earth around 292 million years ago and about 65 million years ago; each of these events (the End-Permian and K-T events, respectively) resulted in mass extinctions and ecosystem disruptions that

---

[1]   Finney and Jones (1985): 335.

[2]   Isaac Asimov reviews a number of possible disasters for humanity in *A Choice of Catastrophes* (Asimov 1979), as does Sir Martin Rees in *Our Final Hour: A Scientist's Warning* (Rees 2003). We review the topic in Chapter 4 of this book.

today would extinguish the human species.[3] Terrestrial threats include nuclear war, pandemics and ecosystem collapses.

Whatever their source, threats to humanity can be classified according to their impact. Some might extinguish our species, others might extinguish a substantial portion of our species, while others might leave much of humanity alive but bring about structural disintegration of civilization, for example, by disrupting the agricultural backbone of civilization. We feel that the disintegration of modern civilization (examined in detail later) would be a vast loss for our species; while modern civilization is beset with problems, many of its own making, we consider it nonetheless worth saving because its benefits outweigh its detriments.

Premise 1, then, argues that there exist real threats to humanity as a species and to modern civilization, and that, in the long run, the only solution is to continue with the exploratory, expanding nature of most successful species, so that migration from the surface of the Earth results in viable extraterrestrial human populations that are ecologically, biologically and culturally independent of Earth.

Premise 2: Human migration into space will be the continuation of the natural processes of evolution.

By this we mean – as we show throughout this book – that humans and our ancestors have a 4-million year legacy of migration, and that there is no logical reason to differentiate ancient migrations across Earth habitats from migration off of Earth. Placing human space migration in the terminology of human adaptation and evolution, we argue, is both appropriate and will be important to both its acceptance and its success.

By 'human', in this book we mean any member of (or the entirety of) the biological species *Homo sapiens* (sometimes the subspecies *sapiens* is also added). Later we will show that anthropology differentiates between anatomical and behavioral modernity when considering what it is to be human; for the moment, though, we simply mean modern humans; any member of our species, *Homo sapiens*.

By 'space', we mean any environment beyond Low Earth Orbit ('LEO',[4] here defined as being less than 2000km from Earth's surface). Space, then, is widely defined here, to include the Earth's Moon; other planets and their moons; orbits around those other planets and moons; space objects such as asteroids; and

---

[3] Palaeontologist David Raup reviews these and other extinction events, pointing out that "...
    no species of complex life has existed for more than a small fraction of the history of life. A
    species duration of ten million years is unusually long ..." but also that "widespread species
    are hard to kill." (Raup 1993): 192–193. We comment on these issues later in this book.

[4] Our definition of LEO is from NASA's Global Change Master directory engineering guide: see
    http://www.gcmd.nasa.gov/User/suppguide.

open-space habitats such as planetary 'Lagrangian libration points' (L-points) where free-floating enclosed habitats (commonly thought of as 'space colonies') can remain in place without orbiting a body (other than the Sun in the case of our Solar System). Having said this, we agree with many authors that Mars is most likely the first off-Earth place that humanity will colonize in substantial numbers.

By 'migration' we mean purposive, permanent movement of humans from Earth to other habitats where they can survive as biologically, materially, philosophically, culturally and politically independent populations. By specifying purposiveness in human migration to space we indicate that this migration must be carefully planned and proactive, rather than hastily planned and reactive, as in the case of an emergency; in Chapter 3 we will introduce the concept of *niche construction* to identify this uniquely-human proaction, which has, ultimately, allowed the survival of our genus and species so far.[5] By specifying permanence in this migration, we are speaking not of 'tethered' explorations planned and carried out to return humans to Earth – though those will of course continue for some time – but 'untethered' movements in which migrants to do not return to Earth. Those humans who do not return to Earth will first be considered colonists, because they will have some dependencies on Earth or near-Earth supply. After some time, however, these humans would be considered entirely new, independent human populations: extraterrestrial populations. They will, naturally, diverge from humans of Earth both biologically and, more so, culturally.

By using the word *adaptation* in this premise, we mean that human migration to space is properly considered adaptation in the same way that ancient human migration into the Arctic region, for example, was a natural process that can be understood at least partly in biological terms as adaptation. Adaptation, as a verb, indicates any adjustment to an environment to increase survival. As a noun, *an* adaptation is a specific biological or cultural structure or process that increases ones' probability of survival.[6] A hummingbird's long, slender beak, for example, is an adaptation to the environment in which it lives, which includes

---

[5] Moving populations of humans to new habitats would have effects on culture as well as biology, and must of course be carefully considered; an evacuation is different from purposive colonization. It has been shown that encountering new conditions can stimulate cultural innovation (Barnett 1953). Regarding biology, migration is one of the most significant drivers of genetic population dynamics and evolution of any species, which is why it is included as a factor in the well-known Hardy–Weinberg model of gene pool equilibrium; see Minkoff (1983): 146.

[6] As we will see, an adaptation is, to paraphrase biogeographer Geerat J. Vermeij, any variation that allows an organism to live and reproduce in an environment where it probably otherwise could not survive (see Vermeij 1978). A little more narrowly, and closer to the level of the individual organism, an adaptation can be considered anything, such as a physical characteristic, that improves fitness.

flowers whose nectar is only available deep inside long, slender tubes of petals. The traditional Arctic ethos, in which injured individuals voluntarily commit suicide during hard times to increase the likelihood of survival of the rest of the group, is an example of a cultural adaptation.

In sum, Premise 2 holds that human migration into space will be a continuation of human evolution rather than an outlandish, science-fiction fantasy of space 'conquest' or a soul-less, technocratic endeavor focused on machines – both common conceptions among those who oppose human space activity. Rather, human space migration will be the same thing migration has been on Earth: people in search of new homes, using both biological and cultural adaptation to survive and, eventually, thrive.

Premise 3: Anthropology – the scientific study of the human species, in all its aspects from biology to culture – will be critical to successful human migration into space.

This premise holds that if humanity is to succeed in migration to extraterrestrial habitats, the endeavor will have to be strongly informed and conditioned by what humanity knows about itself. The study of humanity is in Western science known as anthropology, *anthropo* referring to humanity and *logy* referring to 'the study of'.

Anthropology is the only academic discipline that has systematically studied the human species at large; sociology and psychology take rather narrower views, normally focusing on Western, industrialized cultures, whereas anthropology examines all humanity in space and time. Anthropology is traditionally subdivided into four subfields. *Physical anthropology* has studied the human species as a biological phenomenon, and since biology is structured by the process of evolution, physical anthropology is deeply involved in the study of human evolutionary history. This includes the search for, discovery, and analysis of fossil remains of our ancestors and near-relatives, the study of the entire Primate order (of which *Homo* is one of many genera) and – largely in the last 30 years – the genetics and genome of our species. Another major division of anthropology, *archaeology*, studies the human past via the material remains of human lives; artifacts, from entire ancient cities to individual stone tools that reveal the long-term story of what we and our ancestors did in the past. A third major field, *cultural anthropology*, studies living human groups, referred to as cultures, to understand how humans organize themselves, communicate, engage in conflicts and mediation, and so on. Finally, *linguistic anthropology* studies the unique human communication system called language, from its origins to how it is used among living cultures. In sum, these studies have provided humanity with a Western scientific understanding of its origins, story, and current state.

In Premise 3, then, we argue that if humanity is to succeed in migrating to space, that migration will of course have to be structured on what humanity is, and how humanity changes through time. Since anthropology is the only Western academic field that has studied what humanity is and how it has changed (and continues to change) it will be a critical field in the design and carrying out of successful human space migration.

How Anthropology Will Assist Human Space Migration

What can the four main domains of anthropology offer to assist human space migration? A brief look at what each subfield of anthropology has to offer can help outline its potential contributions.

Physical Anthropology

Physical anthropology, studying humanity as a biological phenomenon with the tools of biology, will be critical to human space migration. From the perspective of individual human beings, physical anthropology is the only field equipped to understand the human body in all its evolutionary detail, from anatomy and physiology in addition to its *bio-cultural* evolution. The human body, and its processes, are not entirely shaped by biology, as we shall see; some of our characteristics evolved in significant tandem with cultural behavior. A long history of investigating the effects of space travel and LEO-type microgravity occupation has resulted in a large corpus of data that warns of dangers to the human body, as well as dispelling some myths–we know that humans can eat and drink in micro- or zero-gravity, that the stresses of reaching orbit and re-entry can be sustained in current-technology vehicles, that long-term exposure to microgravity can result in the weakening of various muscles and so on. But these studies have focused on relatively short-term concerns of individuals, whereas the population genetics and cultural elements of biological evolution at large are the expertise of physical anthropologists, and these will be increasingly important as we consider longer-term exposure of human gene pools and populations of human bodies to extraterrestrial environments.

Especially in the field of human population genetics, physical anthropology will be critical to understanding and maintaining the genetically Minimum Viable Population (MVP)[7] required by the human genome to propagate in health. Currently believed to be around 500 individuals, this number can be investigated and refined by physical anthropologists equipped to understand the genetics of ancient human colonists, such as the Vikings, who colonized southern Greenland, Iceland and Eastern Newfoundland in the first millennium AD. Physical anthropology is decreasingly a field dominated by the study of hominin fossils and comparative primate anatomy and behavior, and increasingly embracing the study of the human genome, human gene function and

---

[7]   Minimum Viable Population differs per organism, but in many cases is remarkably low, on the order of tens to the low hundreds of individuals (Raup 1993): 125–126. We revisit this issue later in this book.

regulation, and comparative genomics – the comparison of genomes between species to better understand the evolutionary process, an understanding that will be of central importance to the human colonization of space. In this and other ways, physical anthropology will be critical to helping structure the migration of humanity off Earth.

Archaeology

The field of archaeology will be important as well. A century and a half, more or less, of modern archaeology has sketched out the essential timeline of human and pre-human history and prehistory. This study has revealed and analyzed the artifacts remaining from the lives of countless migrants and explorers of the human past; humanity first moved out of Africa equipped with stone tools possessed by no other creature on Earth; it then successively occupied one more distant and otherwise-hostile ecological niche after another, including the Pacific and Arctic, equipped with ingenious tools such as sailing craft and snow goggles, respectively. Every artifact related to one of these migrations tells us something about the human capacity for innovation that lies at the heart of our species' ingenuity and survival. Archaeological studies tell us about adaptive and colonizing failures as well, as in the case of the Greenland Norse, or the earliest colonists of Easter Island. Although these people lived in ancient times, and our technologies today are beyond them, materially they *could* have survived, but culturally they enacted courses of action that drove them to extinction. In some cases, then, archaeology shows that in addition to possessing the tools needed to survive, human cultures must value or at least tolerate change and adaptation to new circumstances if they are to survive. This acceptance of change and variation recognizes and aligns humanity with one of the more important realities of the Universe, which is change, as we will see throughout this book.

Archaeology, then, gives us a context for human space colonization. It would also, however, be important in maintaining the heritage of the earliest colonists. When European colonists of the New World arrived, they spent little energy thinking about the distant future, and to understand them today we laboriously excavate the remains of their original settlements because historical records of their time are incomplete. Space colonists knowledgeable about archaeology will take care to arrange things such that their distant descendants will be able to understand their lives, even with futuristic excavations. Equally importantly, archaeology has built a large body of knowledge regarding the human use of objects – technology – to achieve certain objectives over time. While ancient technologies may be out of date specifically, the human use of them to survive is of important philosophical interest, as we will see throughout this book. Archaeology's understanding of the world, then, of *material culture* (introduced below) will also be important to human space colonization.

Cultural Anthropology

The contributions of cultural anthropology to human space migration will also be substantial. Like archaeology, cultural anthropology gives us a context in which to understand migration into space. As we shall see throughout this book, despite a handful of genetic (biological) adaptations to environments worldwide, such as skin hue or body height, most human adaptations over the past 100,000 years have not been physical, or *somatic* (*soma* referring to the body). Rather, the bulk of human adaptation has been *extrasomatic*, that is, non-bodily, or cultural. We are not born knowing how to build an igloo or a seal-hunting kayak, but we can be *enculturated* to know how to build these adaptations to traditional Arctic life, for example, by instruction from our parents and peers.[8] While other species make and use tools, as we will see in this book it is our species that is absolutely reliant on complex, highly-specific tools (as well as cultural behaviors) to survive. Cultural anthropology, then, is critical to understanding one of humanity's most distinctive traits; survival not only by matter, but by mind.

More practically, cultural anthropology will be critical to conceptualizing, implementing, and keeping alive eventual extraterrestrial colonies of humans. Cultural anthropology has a rich understanding of migrant experiences specifically, and cultural evolution at large. Such knowledge could be used to help plan, or at least interpret, the inevitable alterations of human cultures that will take place when humans begin to live independent of Earth. Some argue that specific ideas that might be forwarded to help human space colonization would be 'social engineering', but it is our view that every law we pass, every tradition we take up or drift out of, and every decision we make are already 'social engineering'.

Linguistic Anthropology

The contribution of linguistic anthropology to space migration and colonization will be important, but somewhat indirect. While many life forms communicate with all manner of signalling systems, including visual, audible, and chemical, humanity's linguistic communication system is unique in several ways, and critical to the intelligence that humanity depends upon to survive, and has for over 100,000 years. As we'll see later in this book, the cognitive structures and workings of human language are directly related to and facilitate the capacity we call intelligence. Like culture itself (as just discussed), intelligence rather than biological adaptations have been key to human evolution over the past 100,000 years.[9]

For the first century of anthropology, the origins and evolution of both

---

[8] Biologist Theodosius Dobzhansky (1900–1975) put it this way: "In man, the most important kind of plasticity [for survival] is his educability." See Dobzhanzky (1971): 10.

[9] An approachable review of current ideas regarding the evolution of the mind and language is available in Smith (2006); we review this material later in this book as well.

language and intelligence were so difficult to study that little significant progress was made in understanding these central features of human evolution. Considerable advances in the last few decades, however, have shown that the two are critically linked, stimulating a current revolution in the study of these unique features, as we will see in Chapter 3, *The Adaptive Suite of Genus* Homo: *Cognitive Modernity and Niche Construction*. The role of linguistic anthropology, therefore, in space colonization will not be direct but will be important to continual development of our understanding of what intelligence is, leading to a better understanding of how to use it.

Table 1.1 summarizes some of these points about the significance of anthropology for the planning and carrying out of successful human migration to off-Earth environments.

Anthropology and Space Colonization

Our message here is that the common conception of anthropology being a 'fuzzy' field distant from high-tech endeavors such as space exploration – and, ultimately, space colonization – is a myth, or at least not the entire picture. Certainly some aspects of anthropology are deeply concerned with social behavior that is difficult to quantify (if it is even appropriate to do so): but this is not true of all anthropology, as we will demonstrate throughout this book. And since space migration is about humans finding and adapting to new places to live, it will be a continuation of human adaptation and evolution, and anthropology is the study of humanity and its adaptations and evolution. No other field is so well equipped as anthropology to organize the human element of the colonization of extraterrestrial habitats, but so far in the history of human space exploration, anthropology has not played a significant formal role. The human factors involved have been, as mentioned, the domain of flight physicians, even when long-term space habitation – for several astronauts, now, extending over a year – has been carried out. For example, the bulk of that work has been carried out in LEO, with small crews rather than communities or whole cultures. The same applies to the human lunar exploration program of the late 1960s and early 1970s, which used small crews for very limited periods (not exceeding 14 days) off-Earth.

Such short, exploratory forays, as fantastically exciting and important as they have been, are not the human space activities we refer to in this book. In this book we are talking about viable colonies of thousands of human beings – actual cultures – living out their lives in new habitats off-Earth. These are not meant to be experiments or explorations (though both will continue); they are meant to establish true extraterrestrial human populations that will ultimately culturally, philosophically, linguistically and even biologically diverge from the Earth stock of humanity. Such large-scale – in space and time – concerns cannot be envisioned or understood using the terminology or the cognitive framework related to human space exploration so far. Anthropology is needed.

Table 1.1. Major Fields of Anthropology and their Place in Assisting Human Space Migration.

Field	Study	Methods	Contributions to Space Exploration So Far	Expected Contributions to Human Space Colonization
Physical/Biological	Humanity as a biological species	Biology, evolutionary biology, genetics and genomics	Understanding human tolerances to extraterrestrial regions so far explored	Continuation of current work; focus on human population genetics of colonies; focus on human developmental and population genetics in extraterrestrial environments.
Archaeology	Humanity's past as revealed by artifacts (objects) remaining from past human activities	Excavation and analysis of material culture (artifacts) remaining from ancient cultures	Outlining the multi-million-year history of human adaptation to new environments with many and diverse technologies; providing a human context for human space migration	Understanding technological adaptations of colonists and explorers; understanding history of adaptations; safeguarding extraterrestrial heritage
Cultural Anthropology	Modern human cultural similarities, differences, social organization and social dynamics	Participant observation; analyses from emic (insider's) and etic (outsider's) perspectives	Provides an understanding of human behavioral and cultural variability and social dynamics, and a context for human space colonization	Substantial assistance in the planning, establishment and flourishing of extraterrestrial colonies
Linguistic Anthropology	Human language, including physiology (anatomy) and cognitive mechanisms including grammars and vocabularies	Study of anatomy and physiology of speech, gesture and cognitive bases of language	Provides a critical delineation and understanding of the cognitive tool – language – that allowed modern human expansion and adaptation worldwide	Assistance in understanding the divergence of languages over time, their conservation, and aspects of communication between extraterrestrial populations themselves and Earth

Table 1.2. Significant Differences Between Human Space Exploration To Date (2011) and Envisioned General Human Space Migration and Colonization.

Space Activity Mode	Number of Participants	Duration	Essential Objective
Exploration	1 (e.g. Mercury & early Soyuz) up to 14 (maximum occupancy of International Space Station)	Minutes (early Mercury) to <2 years continuously (Mir and International Space Station occupancy to date) and >20 years discontinuously (in the past two decades there has almost always been at least one human in LEO)	Exploration, gathering information, understanding of off-Earth environments
Colonization	Low hundreds (human Minimum Viable Population on order of 500), to thousands and eventually millions/billions	Decades, centuries and beyond	Establishing new cultures civilizations materially, culturally and biologically independent of Earth

Table 1.3. Associative Shifts Required from Old to New Domains of Thought to Promote Envisioned General Human Space Migration and Colonization.

Old	New
Cold war/military/nationalistic	Humanistic/families in space/people
Industrial, mechanical	Human, organic
"The Right Stuff"/professional	Everyman/Do-it-Yourself
Short-term, limited goals	Long-term, infinite future
Exploration	Colonization
Conquest of Nature	Adaptation to Nature
Limitation of human options	Opening of human options
Hardship/endurance	Challenge/pioneering invigoration
Barely possible	Possible & routine
Technical mystery/inconceivable	Demystified/under-the-hood
Costly luxury	Responsible investment
Destiny/'march-of-progress'	Evolution
Return on investment = spinoff	Return on investment = human survival

To assist in real human space migration, we need to start almost from scratch; we need a cognitive framework of colonization, rather than exploration (see Table 1.2), and we need a new vocabulary. In this book, we argue that that framework and vocabulary will derive from one particular perspective encompassed by and investigated by all the domains of anthropology, the perspective of long-term evolutionary adaptation.

As important to conceiving of human space colonization as fundamentally different from human space activity to date is the need to make cognitive shifts in the associations held by the general public with human space activity.[10] We suggest some of these required shifts in Table 1.3; they will be discussed later in this book, but they are important to introduce here to establish our general tenor.

An Anthropological and Evolutionary Primer for Building an Adaptationist Paradigm for Human Space Migration and Colonization

Tables 1.2 and 1.3 clearly indicate the differences between human space activity so far carried out and what is proposed in this book. We argue that it is not enough to 're-tool' the existing exploratory-mode human space activity into space colonization, but that a new cognitive framework for thinking about space colonization is required.

Further, since human space colonization will be about humanity moving into space (rather than exploring it specifically, and then returning to Earth), that new framework should be informed by the science that has studied humanity over the long term: anthropology. Additionally, the field of evolution, and specifically evolutionary adaptation, will provide critical guidelines not only for implementing human space colonization, but for conceiving it from first principles. In sum, human space migration and colonization should be guided by the anthropological study of human evolution and adaptation.

To begin this paradigmatic shift in our concepts of humans-in-space, some basic terms must be clarified. The rest of this chapter is a brief primer of anthropology and evolution necessary as a background for building an adaptive framework for human space colonization.

Anthropology

As we have already seen, *anthropology* is the field of study encompassing all aspects of human behavior and evolution, and throughout this book we use the word to refer to this academic discipline alone. Anthropology is similar to *sociology*, which studies a variety of aspects of human behavior, but sociology is characterized by its focus on urban and modern populations while anthropology considers the historical and evolutionary context of such populations and uses a global, comparative approach and other methods we will see throughout this book.

---

[10]   The American public has long supported human space activity despite the government's inability after the Moon missions to equal human space flight accomplishments with public expectations; see a review in McCurdy (1997), especially *Prologue*. The same interest is present in Europe (e.g. see Tamm 2006). As human space activity becomes increasingly privatized, it is significant to consider what shifts will be necessary to maintain that interest.

Evolution

Evolution is the process at the heart of the life sciences. For various reasons it has a high public profile but it is commonly misunderstood, so a clear explanation of what we mean by the word in this book is essential.

Evolution in this book refers to the process by which the properties of life forms change over time by three core processes; replication, variation and selection. *Replication* refers to the production of new generations of life by parental generations. This is largely synonymous with reproduction, but the word replication has subtle differences that we discuss in Chapter 2. *Variation* refers to the differences seen in offspring both between one another (siblings and other members of the same species) and their parent generations. While some life forms appear to produce near-clones, genomic study in the past two decades shows this to be largely an illusion – even a single base-pair variation in DNA in three billlion human base-pairs could have significant effects to the human being, for example – and that the rule of replication in nature is that it nearly always produces variation (even if it difficult for humans to observe) rather than clones. *Selection* refers to the differential replicative success of life forms. Because all members of populations are slightly different, not all members of a given generation of life forms have the same likelihood of survival to reproductive age, and even those that do have their own offspring do not all have the same number of offspring. Rather, certain genes are propagated through time, and which genes are propagated depends, often, on environmental conditions, which are always changing in one way or another, resulting in the change of life forms over time, subtly or obviously. These three factual and observable processes of replication, variation, and selection are not unified as a single entity, but they do result in change over time and with human perception a line is drawn around them, labeling their consequence *evolution*.

Those members of a population that are more likely to have more offspring than their contemporaries are called *fitter* than their contemporaries. *Fitness* (a life form's likelihood of having offspring) is largely dependent on the character of the variations it possesses; variations that are beneficial are generally *selected for* and those that are not beneficial are generally *selected against*. Which variations are beneficial and which are not are not archived in some cosmic computer; it depends on the *selective pressures* that 'evaluate' (though without consciousness in most cases) the variations among a given population of life forms. *Selective pressures* are numerous and can change over long and short timescales. For example, the appearance of a predator where moments before it was absent is the introduction of a significant selective pressure that will shortly evaluate the variations (for example, musculature related to predator-avoiding leaps possessed by members of the prey population) in a given population.

Over time, then, as selective pressures change, life forms change as well. Note that they do not do so (except in the case of humanity, which we will come to later) by intent; there is no known way for life forms to alter their own DNA such that their offspring will possess certain beneficial variations. While we increasingly realize that there can be important and heritable environmental influences on the genome, largely we see that the sources of variation (to be discussed later) act in a rather random manner.[11] Kinds of life that do not produce variations sufficient to endure their selective environment become *extinct*, a common evolutionary fate; palaeontologist David Raup has calculated that over 99% of life forms since the origins of Earth life have become extinct. Life forms that do endure changes in selective pressures are considered to be *evolving*, and in fact if replication, variation and selection are occurring (none may be removed from the evolutionary process without terminating it) then we may say that evolution is occurring. Over time, as life forms change according to selective pressures (which might come *to* a given life form, or *be encountered by it during migration into new* areas) they can be said to be *adapting* to such changes.

So far we have used the term *life form* to mean either an individual life form or the properties of a population of life forms. To clarify this (though the use is appropriate in many cases) consider that a *species* is a population of life forms that breed among their own kind, but not with other species. Biology currently recognizes that species are not necessarily easy to delineate, and a spectrum metaphor for life forms is commonly used when speaking at large, but it remains that species are real; elephants and mosquitoes, for example, though sharing some DNA (as do all life forms), breed separately and are thus different species.

New species arise largely when barriers to reproduction arise within a population of a single species; this is called *reproductive isolation* and it may be physical, as when a new river divides a previously single population into different populations on either side of the river, or behavioral, as when a subgroup of a single species population begins to behave differently than the rest, perhaps in relation to new, local rather than range-wide selective pressure. When reproductive isolation persists to the degree that the two previously-interbreeding populations can no longer interbreed, we may say that speciation has occurred.

Finally, the central message of biology is that DNA, the molecule that directs the building of life forms, builds those life forms, which are subject to selective pressures in various environments. In a more abstract sense, the information content of the DNA is considered the *genotype*, and the built life form the *phenotype*. The relationship is shown below; note that in this system information generally moves only one way, from genotype to phenotype to environment; although new research shows that environment can effect DNA, altering its information content, this does not normally occur in an organized way and the message below remains central to understanding evolution;

Genotype → Phenotype → Selective Environment

---

[11]   For a brief but informative overview of the 'Postgenomic Evolutionary Synthesis' see Smith and Ruppell (2011); for details see Koonin (2009a, 2009b).

Though biology is continually refining what we know about evolution,[12] the encapsulation of evolution just presented is sufficient to understand its core processes. Note that sometimes the term *Darwinian* or *neodarwinian* are used to distinguish the modern understanding of evolution from other models of the change of species through time and that it is neodarwinian evolution that is today the foundation of the life sciences.[13]

Figure 1.1 visually summarizes the evolutionary process of replication, variation and selection. In panel (A) we see "Replication" as the mating of two life forms. Under "Variation," we see that while the offspring are largely similar to their parents and one another, two have a variation; the bottom two are shaded. And under "Selection," we see selection; for whatever reason, the darkest-shaded offspring is not well suited to the environment (it is less fit than its companions), and it does not survive long enough to pass on the genes for its dark coloration; it is 'selected against'. In contrast, while the upper two offspring survive, another variation, the medium-gray shaded circle at the bottom, also survives and passes on the genes for that variation. Over time, if that variation is beneficial, it should become common in the population. In the central panel (B), we see a case where two parents mate, having four offspring that are close copies, but with one lighter-colored variation. For whatever reason, that lighter color is deleterious – it reduces fitness, somehow making that individual a little less well-suited to its environment. For this reason, under "Selection," the individual carrying that variation is selected against (X), and the genes for that variation do not continue in the gene pool or will be very rare (in the event it lives long enough to have a few offspring). Finally, in panel (C) we see a case where a lighter variation resulting in the offspring of a pair of replicators is selected for, for whatever reason that lighter variation conferring some kind of advantage. In contrast, the darker-colored offspring, who once were in the majority and were the most fit, might be actually selected against if, for example, the environment changes and being born with characteristics that worked in their parent's time are no longer beneficial. With the selective pressure now changed, the genes for lighter coloration become more common in the population, because they confer a greater fitness. Over time, if reproductive isolation arises – for whatever reason the darker and lighter forms no longer interbreed (dotted line) – the populations might diverge so much that speciation occurs; there are now two kinds, a light and a dark species, whereas before there was only one.

---

[12] The 'Modern Evolutionary Synthesis', crystallized in the 1950s, is currently being overturned by a New Synthesis in which advances in understanding the DNA and the genome have resulted in evolution being newly understood as a result of a "plurality of evolutionary processes and patterns," resulting in an "incomparably more complex" understanding of the world of living things than ever before: for a review see Woese (2004). A brief review of these changes can be found in Smith and Ruppell (2011).

[13] A fuller treatment dispelling common misunderstandings of the evolutionary process is available in Smith and Sullivan (2006).

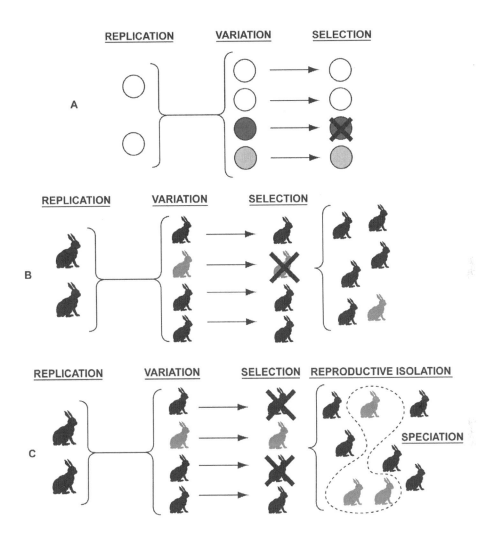

Figure 1.1. Visual summary of the processes of Replication, Variation and Selection that Result in Evolution.

Evolution is so important to the entire message of this book that it is important to dispel some common misunderstandings of evolution circulating in the mass media.

First, evolution is not a coherent system; 'it' does not 'do' anything because the word evolution refers simply to the *consequence* of replication, variation, and selection, processes that are not linked, united, or combined in a coherent, functional whole. It is difficult to overcome the sense that evolution is an engine that 'tinkers' with the shape of life forms or 'punches out' species over time, but this wording gives the evolutionary process an intent that we do not see in nature.

Second, because of this lack of 'itness' or intent we see that evolution does not have a purpose. Contrary to what we might read or watch in mass media, evolution does not 'cause' extinctions or 'create' new species with any kind of coherence or 'attempts' to do anything. Evolution simply occurs.

Third, that evolution simply occurs does not mean that 'evolution is random'. This is again poor wording even though it is found even in some biology textbooks. Randomness does not apply to replication; it simply happens that life forms do have offspring. What variations occur in offspring *are* essentially random, but, critically, which offspring survive to have their own offspring *certainly is not random*; those that are fitter than their contemporaries have more offspring, and this 'pushes' DNA with variations that make for fitness in the present moment into the future. Note that in this respect evolution is *reactive* rather than *proactive*; any life form possesses DNA that was suitably fit for its *parents'* selective environment, but evolution cannot look into the future and tailor its DNA for new selective pressures on the horizon.

Finally, as will already be clear to the reader, human evolution is in many ways different than the evolution of all other life forms. Human consciousness and awareness of the evolutionary process makes human evolution uniquely proactive; we can and do see changes on the horizon, and alter ourselves and our environments to cope with it. That alteration, though, does not occur by significant DNA manipulation (today or in the past 100,000 years) but rather by manipulation of our culture. Culture, we will show in this book, is the information a group of humans share and pass onto their offspring; it is the instruction manual for how to survive and it can be used to build physical objects – tools – to cope with changes in selective pressures. Since culture is what humans use to survive rather than (largely) our biology, and culture *is* eminently mutable, human evolution by cultural adaptation is uniquely proactive as well as very rapid. Other life forms must wait (though they do not even know they are doing it) for beneficial variations to arise if they are to survive new selective pressures, while humans *invent* new cultural behaviors and technologies to cope with change, to adapt. In this way, then, while humanity certainly does evolve – and some of our biological evolution has been and continues to be important – it must be remembered that humanity evolves differently in fundamental ways. That does not mean that evolution does not apply to humanity, however; no life form can avoid replication, variation and selection and their consequences. But in humanity there is in addition to biological (DNA) evolution, *cultural evolution*, the change of the information content of a group of humans that guides their behavior. We will explore this parallel channel of evolution throughout this book.

Adaptation

Adaptation, as a verb, indicates the process by which life forms (again, not consciously) change in reaction to selective environment changes, bettering the 'fit' of the species to its environment over time. As a noun, *an adaptation* is anything that increases fitness, which, according to taste, can refer to the likelihood that an organism will have offspring, or the actual number of offspring a life form has. In either case, such 'anythings' can include physical (phenotypic) characteristics that a life form is born with, such as better or worse eyesight than one's siblings, as well as behavioral variation, some of which can be genetically linked (as in the case of flatworms that either graze alone, or as part of a larger community, based on a single change in one of their genes), or learned in the course of life through imitation or active teaching. Some consider behavior so important that it is given the though-provoking term *extended phenotype*, an extension of the gene, so to speak, in space and time (accomplished by complex behavior), that reminds us of the anthropological conception of culture as the extrasomatic means of adaptation, as mentioned earlier.[14]

We will return to detailed examinations of adaptation later in this book, but for the moment it is sufficient to consider that *adaptation* can refer to a *state* of adaptedness (on a spectrum running from poorly-adapted to well-adapted), a *characteristic* that allows life in a given selective environment (e.g. webbed toes on a seagull) or the *process* by which a life form accumulates the capacity to live in a new ecological niche over time. Regarding individual characteristics, evolutionist Robin Dunbar has succinctly written that:

> "When we speak of adaptation in an evolutionary context, we mean to suggest that a particular trait confers a selective advantage – in other words, that a given trait allows its possessor to contribute more copies of a given gene [used in building that trait] to the next generation than any alternative trait of the same kind [gene]. This is the benchmark against which evolutionary biologists judge a trait's adaptiveness."[15]

Culture

The anthropological term *culture* has been defined many ways in the last 150 years and it remains the subject of active debate. Despite details, however, anthropologists refer to culture generally (though not inaccurately) as a *learned and shared set of ideas that guide some human behavior*. Taking this phrase apart allows us to understand exactly what it means, and why culture is so significant to humanity.[16]

---

[14] The term *extrasomatic* has a long history in biology, from which it was borrowed by anthropology.

[15] Dunbar (1998): 74.

[16] A modern cultural anthropology textbook begins its discussion of culture by quoting Sir Edward Tylor (1832–1917), an early and influential anthropologist, whose definition remains useful, if incomplete: "Culture ... is that complex whole which includes knowledge, belief, arts, morals, law, custom, and any other capabilities and habits acquired by man as a member of society." See Kottak (2009): 27.

By referring to culture as *learned*, we differentiate it from instinct. Instinctual behavior is hard-wired; it is what life forms – from zebras to flies – do based largely on their genetically-determined neural anatomy. Maggots, for example, have an instinctual response to light; they turn away from it until they find shade. How much human behavior is shaped by instinct and how much is shaped by learning – the age-old false choice of 'nature' or 'nurture' – is discussed later in this book. For the moment it is important to note that while some human behavior is instinctual, a great deal of the guides to our behavior are not hard-wired, but learned, from parents, siblings, and peers. We learn by observing and, often, imitating our parents and peers; sometimes we are formally taught. Instinct, then, in a human infant, includes crying out when hungry, while learned behavior (culture) is typically much more complex. For example, as cultural guides to behavior consider the set of navigation instructions that young Polynesian seafarers learn from their mentors. Those navigation instructions are very specific cultural information that is moved from mentor to student according to certain customs, and they are just as important to survival as crying out for nourishment.

One useful analogy for the difference between instinctual and cultural guides to behavior is that of the mind as a computer; in this analogy, instinct is hard-wired, like the most basic operating system that boots up your PC, whereas culture could be considered the specific software loaded onto your PC, as opposed to that loaded onto your neighbor's PC. We'll see later that the human mind is radically more complex than any computer or computer program, but so long as this analogy is not taken too far – as long as it is not taken beyond analogy – it is an illustrative heuristic (thought) device.

Figure 1.2 shows an Afro-Colombian man standing at a temporary shelter in Northwestern Colombia; as much as your mind contains specific cultural ideas about language, music, and appropriate behavior, so does the mind of this man, though those ideas very likely differ to some degree.

Note also that culture is also referred to as being *shared*. Humans live in groups – from families to communities to entire civilizations – and these groups share certain sets of learned ideas that guide behavior. A person born and raised in urban Japan, for example, learns a set of ideas about personal space, gender roles, appropriate foods for certain occasions (and so on) that differ from those learned by a person born and raised in, say, rural Australia. The origins of those Japanese and Australian ideas, and even their specific content (does one eat with chopsticks or a knife and fork?), however fascinating, are for the moment completely beside the point. The point is that groups of humans share sets of ideas that guide behavior. Such groups are often referred to as different *cultures*.

Table 1.4 summarizes the most common domains of the human experience that are guided by specific cultural information. Remember that while every culture has guides to these 'domains' (which tells us very general things about humanity), the specific rules guiding behavior in that domain differ (telling us very specific things about humanity). This is a good list, but some anthropologists would split some of these domains into more than one, and others

Figure 1.2. Afro-Colombian Man in Western Colombia. His mind carries his own culture's concepts of appropriate dress, kinship, food habits, and so on, just as people of every other culture worldwide have such concepts. Photo by Cameron M. Smith.

would add other domains. The table is sufficient, though, for a general understanding of what culture is, with a glimpse at how it varies.

Culture, then, consists of sets of information; these sets can be broadly divided into the above (and other) domains, and they can also be subdivided into their constituent ideas. It is critical to remember that cultural information resides in the minds of individual people; it does not 'ride' on the genes. Culture is acquired not by sexual replication (reproduction), but by social replication, or *enculturation*. Enculturation might be informal, as when a youngster observes and imitates its parents or siblings, or it could be highly formal and structured, as when an Arctic hunter learns about the specific properties of different kinds of sea ice from an elder. In either case, cultural information – sets of ideas – are moved from one mind to another. This is sometimes referred to as *cultural transmission*.

Table 1.4 Common Cultural Domains.

Domain	Concept	Examples
Language	Specific spoken and gestural (bodily) systems of communication, including vocabularies and grammars	Some languages assign gender to nouns, while others do not
Ethics	Concepts of right and wrong, justice, and fairness	Some cultures execute murderers, while others do not
Social Roles	Rights and responsibilities differ by categories such as age (child, adult), gender (man, woman), and status (peasant, king)	Cultures differ in the ages at which people take on certain rights and responsibilities, and specifically what those rights and responsibilities are
The Supernatural	Concepts regarding a Universe considered fundamentally different from daily experience	Different cultures worship different gods, goddesses, and other supernatural entities
Styles of Bodily Decoration	Human identity is often communicated by bodily decoration, either directly on the body or with clothing	Some cultures heavily tattoo the body while others communicate identity more with clothing styles
Family Structure	Concepts of kinship or relations between kin, and associated ideas such as inheritance	Some cultures are polygynous, where males have several wives, and some are polyandrous, where females have several hus bands
Sexual Behavior	Regulation of sexual behavior, including incest rules	Cultures differ in the age at which sexual activity is permitted
Food Preferences	Concepts of what are appropriate food and drink in certain situations	Some cultures eat certain animals while others consider them unfit to eat
Aesthetics	Concepts of ideals (beauty) and their opposites	Visual, musicaland other artistic traditions vary enormously worldwide
Ultimate Sacred Postulates	Central, unquestionable concepts about the nature of reality	Some cultures consider time to be cyclic while others consider it linear

The significance of all this is that one of humanity's most distinctive characteristics is the complexity, richness, and diversity of its cultures worldwide and through time. Some other animals, arguably, transmit social information from parent to offspring, as in chimpanzee societies where mothers seem to very carefully instruct their young in the making and use of tools such as termite-mound probes made from branches, or wooden clubs used to crack open nuts on the surface of a flat rock, used as an anvil.[17]

The point is that while some other life forms do have cultures, those cultures are far less complex than in humanity. For instance, human culture has specific rules for an infinitude of behaviors, such what flag to fly when bringing a boat into a harbor, to how to greet the Queen of England, what gestures are appropriate in a marriage ceremony, and so on.

The chief reason that human culture is so complex compared to that of other primates is that we humans rely on cultural information to survive; we need very specific instructions to navigate not only the physical realities of life – finding and acquiring water, food and nutrients, and keeping ourselves warm, or cool – but to navigate equally-critical social relationships (the wrong gesture at a particular time and place can cost a human their life!). For a number of reasons, humans are very social animals and exclusion from socialization can be intolerable; in ancient Iceland outlaws were often punished by banishing them to the wilderness, totally deprived of social interaction for a set period.

To summarize: *human culture* is a set of ideas that guide behavior. These ideas are socially transmitted from one mind to another; they are learned, rather than genetically inherited. Finally, groups of people sharing certain ideas form *cultures*.

Two caveats: first, while culture is *meant* as a guide to behavior, there is normally a spectrum of adherence to that guidance. Some people adhere very tightly to cultural guides, and some actively disdain them, and many people exist somewhere in-between, and of course many in a culture might share certain political ideals, but differ in their religious traditions. Second, cultures can contain subcultures. Consider 'Deadheads' (enthusiasts of the Grateful Dead rock band) in modern America; are Deadheads 'American'? A visiting German tourist might say "yes, certainly", whereas Deadheads themselves might say "no, absolutely not, we are something else entirely".

Still, the culture concept has its uses, as when we compare the cultures of modern Americans and pioneer Americans, or modern Germans and modern Australians.

However slippery the culture concept can be it has been useful enough that anthropologists use the term in several other ways important to understanding this book.

Archaeologists use of the term *archaeological culture* to refer to *an assumed ancient human culture known by a distinctive set of artifacts or artifact attributes*. Artifacts, generally speaking, are objects made by humans using specific cultural guides. For example, archaeologists use the term 'Lapita culture' to refer to the

---

[17]   Belgian anthropologists Christophe Boesch and Hedwige Boesch-Achermann claim even to have observed a female chimpanzee correct an error in her offspring's use of a nut-cracking club, which they call an instance of active teaching rather than simple imitation; see the video at http://www.youtube.com/watch?v=AElmAJH2G00.

ancient Polynesian people who made and used distinctive pottery, fish hooks, seagoing voyaging canoes and other artifacts dated to about 3000 years ago. 'Lapita culture', an archaeologist might say, 'spread through the Pacific' as evidenced by excavations revealing the spread of Lapita culture from island to island and dating these distinctive sets of artifacts as they appear earliest in Western Polynesia and later in Eastern Polynesia.

Anthropologists also commonly use the word culture in the term *material culture*, referring to objects made or modified by humans. This term recognizes the fact that when people make or modify things, they do so in a way that is informed by their culture. For example, when indigenous Saami people (reindeer-herders of the European Arctic) build a home, they build a structure that shelters them from the wind, insulates them from outside temperatures, and keeps them from laying directly on the frozen ground. Native Inuit people of the Canadian Arctic also build shelters that serve these exact technical functions, but because of many differences between Saami and Inuit lives, environments and cultures, their *material culture* (in this case, shelters) differ. The study of material culture is important to understanding both modern and ancient people, and it's obviously central to archaeologists who study the material remains of ancient cultures – because that is the study of ancient *material culture*.

Finally, culture is often used by anthropologists in the term *culture change*. This refers, simply enough, to the change, over time, of cultures, both modern and ancient. Since culture is a shared set of ideas, culture change is essentially the change of ideas shared by a group of people. For instance, before World War II, American women were not a large part of the industrial workforce, but during the war they began to work in armament (and other) factories. Therefore, cultural ideas about women's work roles changed between the pre-war and the war years: culture change. Culture change is a deceptively simple term; models of culture change – anthropological ideas about the exact mechanisms of cultural change, about what drives it, and how it plays out over time – are many and complex. We will revisit culture change throughout this book.

Table 1.5 summarizes the main uses of the word culture in anthropology and in this book.

Note that modern scientific cultural anthropologists consider culture to be an evolving information set very similar to evolving genetic information sets. In biological evolution, it is the genes in a gene pool that are replicated, vary, and are selected upon, whereas in cultural evolution it is *memes* – information analogues of genes – that are replicated (culturally rather than sexually), vary, and are subject to selection. We will cover this more thoroughly later in this book; for the moment it is interesting to compare biological and cultural evolution in Table 1.6.

Table 1.5. Uses of the Word Culture by Anthropologists and in this Book.

Use	Definition	Example	Notes
Culture as Information	Learned, shared guides to behavior	Guides to sexual behavior; beliefs about the supernatural; specific languages	At root, culture is information that resides in the minds of individual people; it is passed from person to person by many systems of communication, including speech, writing, dance and gesture
Culture as Identity	A group of people sharing certain learned guides to behavior	English culture, Japanese culture, Maori culture, Aztec culture	While it can be hard to draw lines around a culture, and cultures can contain subcultures, on other scales cultures as different as Alaskan native and urban German (for example) can be discerned
Archaeological Culture	A presumed group of ancient people represented by specific characteristics evident in sets of artifacts (see 'Material Culture', below)	'Thule' culture of the Alaskan and Canadian Arctic, dated to over 1000 years ago, represented by distinctive artifacts including bone snow-goggles, certain harpooning equipment for hunting sea mammals, and dog sleds	Archaeologists are careful not to 'reify' archaeological cultures; reification is thinking of an abstract concept as real. Some archaeological cultures are easier to discern from others, just as in living cultures
Material Culture	Objects made or used by humans (artifacts)	The material culture of ancient Mediterranean seafaring includes the use of boats made of wooden planks, whereas the material culture of Southeast Asian seafaring includes the use of boats made of bamboo lashed together	Cultural information – shared by groups that have a common identity – guides differences in how groups produce material culture
Culture Change	Change of ideas (guides to behavior) shared by groups	Early Icelandic (Viking) culture was polytheistic, with many Scandinavian gods, but in 999 AD the Icelanders converted to Christianity	Culture change can be fast, slow, or anywhere in-between, and it might be easy, difficult or impossible to detect in material culture

Table 1.6. Features of Biological and Cultural Evolution Involved in Human Adaptation.

Mode of Evolution	Replication	Variation	Selection
Biological	DNA is arranged in genes, specific base-pair sequences that code for certain protein assembly and other functions; genes are passed from parent to offspring by sexual or asexual *reproduction*	DNA mutation, failure of error-repair mechanisms, and recombination (shuffling of parental DNA) all lead to phenotypic variation in the gene pool	Phenotypic variations lead to fitness differences which lead to some phenotypes having more offspring, which carry the DNA of their parent generation. Selection usually results in a better fit of the life form, over time, to its selective environment
Cultural	Cultural information is arranged in memes, concepts or ideas that specify conditions of the Universe and provide guides to behavior; memes are passed from parent to offspring by enculturation, or *social reproduction*, and memes can be rapidly shared among members of the same generation	Social communication of memes is imperfect, with transmitter potentially altering meme for their own purpose and receiver potentially misunderstanding it; meme innovation is promoted by cognitive structures related to intelligence and language, leading to meme variation in the meme pool (culture)	Memetic variations are 'evaluated' (selected for or against) by social mechanisms (e.g. propaganda or censorship) used to control frequency of memes (ideas) in the meme pool (culture)

Humanity

The term 'humanity' refers variously to modern humans, humans of the future, and some of our ancestors. Anthropologists distinguish broadly between two main kinds of *modernity* in humanity.

Anatomical modernity is easy enough to define, or has been until recently. This term generally refers to possessing a skeletal structure indistinguishable from modern humans. That is, if we cannot distinguish between an excavated, ancient skeleton, and the statistical range of variation we see in modern human skeletons, we call the ancient skeleton 'anatomically modern'.[18] Most anthro-

---

[18]   In the past two decades it has become possible to recover, analyze and compare the DNA of non-human hominins, such as Neanderthals. Current estimates suggest that most DNA will have degraded after 1 million or so years (see Pääbo 2004) but it is likely that older DNA will occasionally be found in other samples

pologists agree that by 100,000 years ago anatomically modern humans evolved in Africa; we will consider their evolution later in this book.

The other sense of humanness is *behavioral modernity*. This refers to whether the ancient humans we are studying *behaved* in ways that are indistinguishable from modern humans. Such behavior includes, at least, complex *symbolism, language*, and *tool use*. Complex symbolism refers to the use of one thing to refer to something else. A stop sign, for example, signals to us to stop driving; it is a complex symbol because the red octagonal shape of the sign does not actually show us a car stopping, we have simply agreed (in some cultures) that this arbitrary shape and color combination mean 'stop the car'. *Complex language*, here, refers to the uniquely human communication system that is structured by complex rules, or *grammar*, and *complex tool use*, here, refers to making and using complex items of material culture, including, for example, igloos, Polynesian sailing vessels, and the Egyptian pyramids. While other animals use some kinds of symbols, systems of communication, and tools, it is in humanity that these are most elaborate, and it is in humanity that they are absolutely required for survival. Using them, broadly speaking, means behaving in a modern way.

Behavioral modernity can be hard to identify archaeologically because it might not leave behind artifacts for archaeologists to study, but for the moment it is sufficient to say that several lines of evidence indicate that, as with anatomical modernity, it is present by about 100,000 years ago, also first in Africa.

When we speak of humanity, then, keep in mind that we might mean anatomical modernity or behavioral modernity; the context of the discussion should make it clear which.

Synthesis

The anthropological and evolutionary concepts above can be parlayed into a new framework for conceiving and carrying out human space migration and colonization; in fact, they must, because human space migration and colonization is going to be about humanity migrating to new environments.

In thinking about this migration rather generally, consider the following characterization, which gives a context for human space migration using the terms introduced in his chapter:

- Humanity is one of the roughly 220 species of primates, relatively large mammals with a 65-million year **evolutionary history** characterized by biological investment in vision as the primary sense and behavioral investment in intensive sociality and intelligence.
- Roughly six million years ago we see the appearance of **hominins** in the primate order, characterized by relatively large body size and locomotion on two feet (bipedalism). By two million years ago some of these hominins are making and using stone **tools** of greater complexity than most other

animal tools; they increasingly compete directly with top carnivores in their **selective environment** of the African savannah, as their increasing brain size demands greater caloric intake facilitated by tool use to process high-calorie animal carcasses.

- Reliance on tools to survive begins a long-term trend in hominin evolution of the **decoupling of behavior from anatomy**: using tools to adapt to changing selective pressures rather than relying on biology alone. By one million years ago this unique mode of **adaptation** allows the **genus** Homo to occupy all of Africa, parts of Europe and all of Asia South of the Himalaya.

- By 100,000 years ago the **grammatical complexity** of **communication** in Homo was unique in the natural world, leading to **cognitive revolutions** including the appearance of modern intelligence, which was used to organize and manipulate ever-increasing storehouses of **cultural knowledge**. By 10,000 years ago, tool use and cultural adaptations were so effective as means of adaptation that Homo had **migrated** to and occupied nearly every conceivable Earth **habitat**, from Siberia to the central Amazon, South America, and parts of the high Arctic, and not just by passive reaction, but also by proaction.

- By 2000 years ago, Western civilization, underwritten by an agricultural mode of subsistence that produced large food surpluses, was breaking from mythological interpretations and explanations for itself and the Universe and devising scientific principles of knowledge generation. By 40 years ago, technological knowledge allowed small numbers of humans to physically distance themselves from Earth and explore the surface of an extraterrestrial body, the Moon.

- By 2010, humanity possessed the knowledge and material necessary to adapt – with culture and tools – to a variety of **extraterrestrial habitats**.

- In 2040 the first human was born on Mars and by 2100 the Martian population was so culturally and genetically distinctive from Earth populations that it was designated as a new subspecies, *Homo sapiens marsii*.

- By 2300 thorough understanding of the human **genome** and cultural acceptance of multigenerational voyages were combined with revolutionary understanding in physics that allowed near-light-speed propulsion. This resulted in launch of the first **transgenerational** interstellar colony towards a planetary system, 20 light-years distant, known by astronomers and long-distance probes to possess habitable planets.

- By 2500 the human populations living on the surface of this first-colonized extraterrestrial planet, as well as their companions who chose to continue to live in the starship, are so genetically so distant from Earth populations, after generations of **reproductive isolation**, that they are a new species: *Homo extraterrestrialis*.

The scheme outlined above sets human space colonization into an adaptive, evolutionary and natural context, which is much of the focus of this book.

With our premises and anthropological and evolutionary foundations in place, we can go on to the next chapter to examine just how it is humanity came to be where we are today – for better and worse, and both literally and figuratively – providing a context for humanity at present.

2 Stardust: The Origins of Life, Evolution and Adaptation

"The ash of stellar alchemy was now emerging into consciousness ... We have begun to contemplate our origins, starstuff, pondering the stars."

Carl Sagan[1]

"The revolutions of Earth are marked therefore not only by physical and chemical changes to the Earth environment, but also by increases in the complexity of organisms living in it. They are marked by increasing use of energy, increased efficiency in recycling materials, and increased information processing [capacity] ..."

Tim Lenton and Andrew Watson [2]

If we are to understand human space colonization as a continuation of fundamentally evolutionary processes, then we have to understand the origin of life – that which evolves – itself; and to understand life, we need an understanding of the origins of the Solar System and the Universe from which the building blocks of life are ultimately assembled. In this chapter, we cover a large span, from the origin of the Universe to the origin of Earth and major events in Earth life evolution so far, including the evolution of humanity's unique, grammar-directed language and consciousness in our genus, *Homo*. We then comment on how knowledge of this evolutionary context can help us best plan human space colonization as an adaptive endeavor.

Origins of the Universe and the Solar System

The human story really began almost 13.75 billion years ago, with the origins of the Universe. In 1931, Belgian physicist Georges Lemaître (1894–1966) proposed a theory of the origin of the Universe that served as the foundations for "big bang" theory, which – in an updated form resulting from advances in astronomy, cosmology and mathematics – is generally accepted today.

Our world, then, and everything on it, is made of atoms that were formed in

---

[1] Sagan (1980):338, 345.
[2] Lenton and Watson (2011): 16.

the nuclear furnaces of the earliest stars billions of years ago, and our Solar System is one of countless planetary systems in the Universe. They share the characteristics of a star (or stars) being the focal point of orbit for a set of planets, moons and other objects such as asteroids and comets.

Formation of the Solar System and the Earth

Modern astronomy recognizes two main generally-accepted models of planet formation: the *core accretion* model and the *disk instability* model. In core accretion, planets form by the gradual accumulation of particles and debris (cumulatively known as *planetesimals*) that gravitate towards one another, resulting in largely solid planets, including Earth-like worlds. In the disk instability model, areas of the protoplanetary disk (a concentration of gas) orbiting a young star accumulate until a critical mass is built, which then rapidly collapses (with gravitational force) into the young star, resulting in gas giant planets, such as Jupiter. Currently, about 1 in 16 known Sun-like stars are orbited by one or more Jupiter-sized gas giant planets, and the ratio increases to 1 in 6 if we include Class A (white or blue-white) stars, which are about twice the size of the sun. Rocky planets are also known to orbit extrasolar suns: in June, 2011, UC Berkeley and UC San Francisco astronomers revealed that at least one rocky planet orbits the double star 55 Cancri (which is about 41 light years from Earth). Almost three quarters of the planet's orbit is in the 'life zone' in which water would remain liquid, and the surface temperature would range from –52°C (–61°F) to +28°C (82°F). Figure 2.1 is a Hubble Space Telescope image of a 'dusty disk' of protoplanetary matter in the Orion nebula.

While our Solar System's planets formed by such processes at various times in the history of the Solar System, our best evidence to date indicates that the Earth was formed on the order of 4.56 billion years ago by the core accretion process. The time from 4.56 to 3.8 billion years ago is known as the *Hadean* era, a term that evokes the hellish conditions of early Earth, characterized by a Sun that shone with only about 70% of its current luminosity, when the atmosphere was mostly water vapor, nitrogen, and carbon dioxide, and when the 'surface' of the planet was molten, dissipating heat from the violent accretionary process; finally, until around 3.8 billion years ago, the new planet was under heavy bombardment from residual planetesimals, splashing into oceans of lava. Not long after, however, the Earth began to cool and the lighter elements rose, forming the mantle and the crust, while the heavier elements, such as iron, settled into the solid iron core, currently sheathed by the liquid iron core.[3]

---

[3]   Eales (2009).

Figure 2.1. Hubble Space Telescope of Dust in the Orion Nebula. NASA image.

We have evidence that Earth's Moon formed just (but clearly) over 4.5 billion years ago – not long after the formation of the Earth. While other known planetary moons appear to be space objects captured by a planet's gravity field, our Moon appears to have derived from the Earth itself. Good evidence suggests that over 4.5 billion years ago the forming Earth was struck by a planetoid slightly smaller than Mars, the impact ejecting molten rock from Earth into space, where it aggregated to form a body – the largest moon (about 25% the size of the Earth) of the inner Solar System – that was captured by Earth's gravitational field. This theory is strongly supported by the fact that the Moon has few *volatiles*, compounds that boil at low temperature, apparently because they were boiled off during the Moon's time as a colossal, agglomerating blob of molten rock.[4]

The formation of the Moon was not an event divorced from Earth's formation, however, because as a result of the collision the Earth's atmosphere was thick with vaporized rock and steam. To an outside observer, Earth at this time might well have resembled Venus, or Saturn's moon, Titan, which has pools of liquid methane but also sheets of water ice, and dense methane clouds.

It is remarkable how short a time, geologically speaking, passed between the settling of the Earth into a relatively solid and stable state and the first evidence for life on our planet.

The Origins of Earth Life

Earth may be the home of life in our Solar System, having evolved independently on our planet, or life may have evolved elsewhere before coming to Earth. In either case, the origins of life itself are of interest, because it is life itself that evolves, and because we are living things.

Independent Evolution of Life on Earth

How life came to be, apparently, from non-life, is one of the largest of all questions in Western civilization, as well as a topic of endless fascination in other cultures. In the scientific perspective, while there is of course some debate regarding the definition of life, most would agree that life is an arrangement of matter possessing three key qualities. First, life forms *metabolize* energy; this is done by some form of digestion, in which larger units of the energy source (food) are broken down – sometimes mechanically, as in chewing, and sometimes chemically, as with stomach acids – to facilitate their distribution of energy throughout the life form's vital systems. Second, life forms are *replicators* that initiate the construction of high-fidelity copies of themselves. This replication is facilitated by molecules arranged in a way that directs the building of the high-fidelity copy (offspring) of the parent generation. Finally, living things have *boundaries* that segregate them from the environment they evolve in as well as from other living things.[5] A somewhat fuller conception of life, compiled from a number of sources, recognizes these and other properties, and refers us to things a little closer to the common conception of life:

- Life forms metabolize food.
- Life forms move, converting some energy into motion (though not

---

[4]   Eales (2009): 154.
[5]   Szathmary and Fernando (2011): 301.

necessarily in locomotion, but any kind of movement, such as a plant's stalk changing shape subtly during daylight and dark periods).
- Life forms respond to stimuli.
- Life forms possess a boundary (e.g. cell wall or skin).
- Life forms change in the course of their own lives (e.g. growth).
- Life forms coexist among others in a community or ecosystem.
- Life forms reproduce, such that similar offspring are derived from parent generations.[6]

How such arrangements of energy and matter formed early in Earth history is a thrilling field of scientific research, and has yet to be entirely resolved. The answers lie in physics and chemistry, because the earliest life was very small compared to what we see of life today in animals and plants, for example (see below), and because early life possessed many important similarities to natural *autocatalytical* compounds that, in the correct chemical conditions, sponta- neously assemble into structures. Recently, German biologist Henry Strasdeit has suggested a three-step sequence (illustrated in Figure 2.2) in which, because of their size and other properties, only certain chemicals of the early Earth environment (which included many organic compounds) entered into compart- ments (hollow chemical structures) where their interactions with other chemicals facilitated change within the system;[7] in this model, the earliest life forms' boundaries are chemical structures that segregated chemical reactions – which became metabolism – inside a shell that might later split, but would then be chemically 'repaired' (again through autocatalysis) such that there would now be two rather than just one life form, each similar to the other because of their common origin. These steps are relatively simple from a chemical perspective, and when 'non-life' crossed the boundary into 'living' things, Strasdeit argues, might not actually be the most interesting question to ask about this issue:

> "Should the self-sustaining and self-reproducing chemical systems of the type described above be considered alive? The answer, of course, depends on how we define life. As the transition between non-living and living matter is gradual, every fixation of a sharp boundary must be artificial. Consequently, there is no consensus definition of life [and] It has even been questioned whether this issue is of major interest at all ..."[8]

The evolution of the molecules that directed more detailed replication of such structures – molecules that would eventually become RNA and DNA – is of course a target of modern molecular research, some of it focusing on the fascinating discovery that living structures are characterized by *homochirality*, the phenomenon

---

[6] This list is compiled from Ruiz-Mirazo, Pereto and Moreno (2004) and Minkoff (1983).
[7] Strasdeit (2010).
[8] Strasdeit (2010): 113.

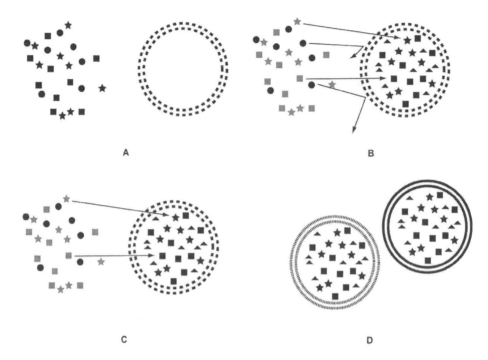

Figure 2.2. Early Life Compartmentalization. In (A) chemicals (stars and other shapes) exist outside natural enclosures, such as bubbles, seen as the dotted circle. In (B), some of these chemicals, due to their size and shape, are admitted into natural enclosures, while others are not. In (C) chemical reactions inside the enclosure lead to early life, which begins to metabolize nutrients (stars). In (D), the boundary of the natural enclosure becomes 'formalized' in the replication of the life form, building the first cellular walls. Image by Cameron M. Smith after Figure 2 of Strasdeit (2010).

in which amino acid molecules twist to the left and sugars twist to the right[9] and how this conditions the interactions of these substances in ways that produce order. In fact, the origin of order, as we have learned in the last few decades, is currently also of great interest, as science itself has drifted away from a conception of order in the Universe as resulting from some kind of divine influence – the perspective of 'Natural Philosophy' – towards recognizing that order can emerge as a property of nonlinear (*stochastic*) and even chaotic processes. Scripps Institute chemist Donna Blackmond concludes her considera-tion of early life evolution by indicating the importance of the field of complexity studies:

---

[9]   Blackmond (2010).

"The pathway to life may be seen as a saga of increasing chemical and physical activity ... The modern field of "systems chemistry" ... seeks to understand the chemical roots of biological organization by studying the emergence of system properties that may be different from those showed individually by the components in isolation. The implications of the single chirality of biological molecules may be viewed in this context of complexity. Whether or not we will ever know how this property developed in the living systems represented on Earth today, studies of how single chirality might have emerged will aid us in understanding the much larger question of how life might have, and might again, emerge as a complex system."[10]

When exactly the transition from non-living chemical systems to living chemical systems took place is also of interest, for which we have better evidence. Though a topic of great internal debate within biology, from some distance it is clearly possible to say that life appeared not too long after the formation of the Earth. With the Earth accreting around 4.6 billion years ago and heavy bombardment of the new planet by planetesimals ending by 3.8 billion years ago, by 3.5 billion years ago there is widely-accepted evidence for life on Earth.

Figure 2.3 shows an astounding case of two fossilized bacteria – one budding off from the other – in 3-billion year-old fossil deposits in South Africa investigated by Harvard micropalaeontologist Andrew Knoll; in this astounding image, the width of even a single human hair would entirely block out the image frame. These early life forms, possessing a boundary, exhibiting evidence of metabolism, and being produced by molecularly-directed replication, were small, on the order of a micron in size (less than 1/100 the width of a human hair), and only much later would larger and even multicellular life forms evolve.

Panspermia

It is also possible that early Earth life was not 'Earthian' at all, but that it arrived on this planet from somewhere in space. The theory of *panspermia*, first associated with Swedish physicist Svante Arrhenius (1859–1927), holds that life has evolved in other places than on the Earth, and that sometimes elements of that life is picked up on certain planets, such as Mars and Earth. For a long time this concept was essentially marginalized but it now enjoys at least wide interest, and has in fact been partly responsible for the development of modern exobiology.

This general interest and acceptance comes from the accumulated knowledge of science. Life, we have found in the past few decades, seems capable of flourishing nearly anywhere; deep sea and other 'marginal habitat' exploration

---

[10]   Blackmond (2011): 16.

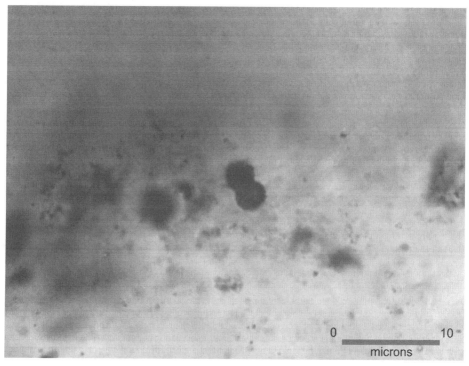

0 ────────── 10

microns

Figure 2.3. Fossilized Early Life Forms in the Act of Replication. Microphotograph courtesy of Dr Andy Knoll, Harvard University.

has revealed life forms that do not derive energy from sunlight, others that metabolize inorganic substances as food, yet others that live at extremely high and low temperatures and pressures, and every condition, it seems, in-between.[11]

Considering this, scientists have returned to principles of panspermia to consider whether life – as we understand it – could survive the three main components of a planet-to-planet transfer through open space, from a *donor* planet or other solid body: (a) acceleration of the organism to planetary escape velocities, involving high temperatures, g-loads and pressures; (b) transit through space, involving zero-g conditions, low temperatures, little or no atmospheric pressure (vacuum), cosmic rays and solar radiation; and, (c) landing (on another planet or solid body, the *recipient*), again involving high g-loads, temperatures and pressures.

The answers, so far, support the principles of panspermia: tests and simulations have demonstrated that bacterial (and other) life can survive all of

---

[11] A good introduction is found in Horikoshi and Grant (1998); updates can be found at NASA's 'Extremophile Hunt' website at http://science.nasa.gov/science-news/science-at-nasa/2008/07feb_cloroxlake/.

these forces, including high g-loads, temperatures down to 10K (–441.7°F) and exposure to vacuum for – so far – a documented time of up to six years. Protected by substances such as amber or salt, furthermore, it has been found that some life forms can be revived after tens of millions of years.[12] Other studies have shown that micrometeorites (on which space-derived life might ride between planets) only a few microns in size can enter the Earth's atmosphere without heating to over 100°C[13] and that during early Earth times, organic molecules freely floating in the Solar System rained on the Earth and other solid bodies. Finally, it is widely accepted that large bodies over 1km in size could be blasted from planetary surfaces as a result of impact by space debris, providing the mechanism responsible for bringing over 40 (known to date) Mars-derived meteorites a quarter of a billion miles, from that planet to Earth.[13] A glance at the conditions of Earth-orbital and interplanetary space (Table 2.1) hint at some of the conditions pertinent to panspermia theory:

Table 2.1 The Environment in Earth Orbit and of Interplanetary Space. Adapted from Horneck et al. (2002):61.

Parameter	Earth Orbit (<500km)	Interplanetary
Space Vacuum Pressure (Pa) (1 Pa=.0001 psi)	10^{-6} to 10^{-4}	10^{-14}
Residual gas (part/cm^3)	10^5H, 2×10^6 He, 10^5N, 3×10^7O	1 H
Solar Electromagnetic Radiation Irradiance (Watts/m^2) Spectral Range (nm)	1360 From 2×10^{-12} to 10^2m	Depends on distance from star From 2×10^{-12} to 10^2m
Cosmic Ionizing Radiation Dose (Gy/year) (1Gy=100 rad) Spectral Range (nm)	0.1-3000 From 2×10^{-12} to 10^2m	=<0.25 depending on shielding From 2×10^{-12} to 10^2m
Temperature Ambient temperature (K)	100–400 depending on orientation to Sun (-280°F to 260°F)	>4 (>–452°F)
Microgravity G force (1g = Earth surface)	10^{-3} to 10^{-6}	<10^{-6}

---

[12]   Horneck et al. (2002): 58.
[13]   Horneck et al. (2002): 69.
[14]   Horneck et al. (2002): 57.

Of known Earth life, the best analogue for panspermic travellers would be the *spore*, a seed-like package of replicating DNA that can survive tremendous deprivation relative to other living things. The spore is a bacterial cell's dormant stage, very resistant to such conditions as found in Earth orbit and even, perhaps, interstellar space (see Table 2.1). The hardy spores of the *Bacillus* bacteria owe their firm constitution to a dehydrated mineralized core enclosed in a thick protective envelope, as well as saturation of the DNA with acid-soluble proteins whose binding greatly alters the chemical and enzymatic reactivity of the DNA. Three space missions have tested survival of *Bacillus* spores, which on the Long Duration Exposure Facility – an experimental platform that orbited the Earth from 1986–1990 – had a 69% survival rate after 10 days exposed to space conditions, 32% survival after 327 days and 1% survival after six years. However, a 67% survival in space rate was found over long periods when spores were shielded in protective sugars or salt crystals (which does occur in nature) and were covered with UV-deflecting foil. Still, cells underwent dramatic restructuring at many levels when exposed to vacuum, including lipid membrane phase changes, irreversible polymerization, desiccation, changes to selective membrane permeability and enzyme activity and DNA alteration; these would not necessarily destroy the spore's potential for life, but could alter it significantly.[15] The alteration of DNA in these space-exposed spores was characterized by both base changes and strand breaks, and such alteration would accumulate in space over time – probably resulting in problems down the line – as DNA repair mechanisms (discussed later) are not active during spore dormancy, and because most calculations suggest planet–planet transfer (as we understand it, in our Solar System) would normally last thousands or millions of years. Despite the hardiness of spores, other problems were revealed by the LDEF, such as the finding that extraterrestrial solar UV radiation kills (in seconds, 99% of *Bacillus subtilis* spores) that space vacuum increases the UV sensitivity of spores, and that mutation rates increase in unshielded space environments.

Despite all of *this*, however, it has been suggested that even after 25 million years in space, most spores would survive if shielded from UV radiation by even just a thin layer of sediment, say 1m–3m thick; and we should keep in mind that Mars belt asteroid *Lutetia* (for example) has a dust layer 600m (nearly half a mile) thick. Finally, we should remember that non-planetary solar system entities are not static, and that over time their suitability as habitats for different kinds of life might be a function of long-term or short-term conditions; comet Lee, for example, was recently observed to produce different molecules over a four-month period as it approached and then shot away from the Sun.[16] In a study published in 1996 it was estimated that spore-like life forms could probably

---

[15]   Horneck et al. (2002): 62.
[16]   Biver et al. (2000).

travel distances of up to 20 light-years with a good likelihood of surviving arrival on a distant planet.[17]

Whether spore or other variety of life or dormant life, it seems plausible – or at least well beyond science fiction – that some kind of life might have arrived on Earth early in our planet's history, and that we, in fact, are 'aliens' to Earth. Until more evidence – such as meteorites bearing evidence of life, or the discovery of life or traces of ancient life on Mars – are developed, however, panspermia will probably remain on the edge of mainstream biology.

The Major Transitions in the Evolution of Earth Life

While the essential elements of evolution are, as we saw in chapter 1, straightforward, it is important to remember that the evolutionary process itself has changed over time. Below, we review the history of Earth life from the perspective of several major transitions, first outlined by British theoretical biologist John Maynard Smith (1920–2004) and Hungarian theoretical biologist Eors Szathmary (b.1959).[18] Recent review of these transitions finds that they remain largely viable in the light of very new genomic and other molecular studies, and that while updates are of course necessary, they outline some of the more important events in the evolution of Earth life.[19]

Figures 2.4 and 2.5 provide an outline the history of life on Earth.

Transition 1: Replicating Molecules to Populations of Molecules in Compartments

Self-replicating molecules – those that will, in the correct chemical environment, serve as a template for producing self-similar copies – are largely known on Earth in the form of DNA, deoxyribonucleic acid, the long, double-helix-shaped molecule first described in detail by Francis Crick (1916–2004) and James Watson (b.1928) in 1953.[20] How the DNA molecule 'came to be' is a study in chemical evolution and autocatalytic processes mentioned above, and at present their origins remain obscure, but the essential transformation is one involving the evolution of information. If information is an arrangement of matter that reflects (and may direct) the arrangement of other matter,[21] then DNA is a molecule that contains a tremendous amount of information, and whose information content is used in self-replication in the correct chemical environment. Earlier information-rich molecules, including RNA according to

---

[17] Secker, Lepcock and Wesson (1996).
[18] Smith and Szathmary (1995).
[19] See, for example, Calcott and Sterelny (2011).
[20] For a review of Watson and Crick's discovery, see Watson and Berry (2003).
[21] For this definition of information as it relates to biology, see Guttman et al. (1982).

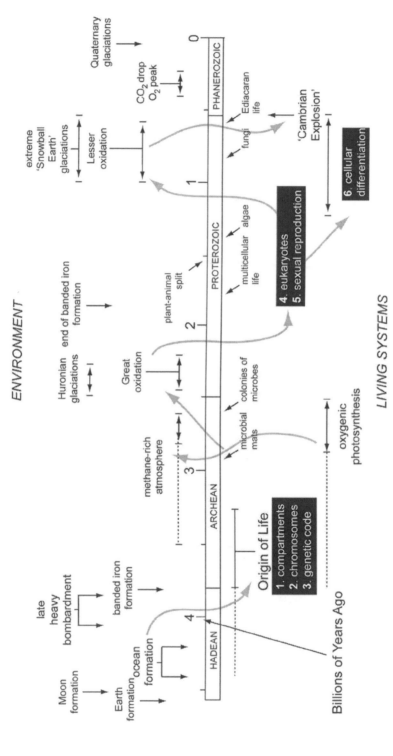

Figure 2.4. History of Earth Life, Part I. The image, adapted from Lenton and Watson (2011), shows the interaction of the environment (upper panel) and living systems (lower panel) over time, shown in billions of years ago on the horizontal scale. The first six main transitions in evolution, mentioned in the text, are identified. Image by Cameron M. Smith.

many evolutionary biologists, were simpler molecules that, over time, accumulated information in molecular form through chemical processes. Precisely how this occurred is beyond our scope here, the essential points being that (a) we need not look to the supernatural to explain these structures, but that they are modeled and understood today in terms of chemistry, and that (b) the evolution of DNA is not widely understood as inevitable – the beginning of some unilineal climbing of a ladder of evolution, from slime to humanity – but a chemical process dependent on chemical conditions.

Where such chemical evolution took place, producing the first Earth-based life (whether or not it first formed on the Earth, or, possibly, on Mars, as suggested by a number of recent studies) is also a point of debate: Henry Strasdeit (mentioned above) suggests that:

> "There are good reasons to believe that volcanic islands were important places for chemical evolution. For example, volcanic lightning in ash-gas eruption clouds very probably produced amino acids ... These compounds, which also came from ... meteorites, dissolved in the seawater ... [but] When lava [subsequently] flowed into the ocean, the seawater evaporated. Subsequently, the remaining sea salt crusts experienced temperatures up to some hundred degrees ... [providing energy for chemical evolution] ..."[22]

Wherever chemical evolution took place with the result of DNA, the fundamental transformation here involved the evolution of a molecule rich in information and which, under certain chemical conditions, spontaneously generated copies of itself. Note, of course, that when we say 'generated copies of itself', no intent or consciousness is implied because, as far as such phenomena are currently understood, they appear absolutely absent in living things for a very long time, a topic we return to later in this chapter when examining the evolution of consciousness in life forms. Earlier in this chapter we discussed the possible origins of the cellular compartments that 'housed' early self-replicating molecules, schematically illustrated in Figure 2.2. Prominent microbiologist Carl Woese has indicated that the development of the cellular boundary was of fundamental importance to evolution, as it sequestered sets of replicators from other sets, leading to a kind of stability of one set as opposed to others, which eventually led to the differentiation of species. It also allowed protection for ever-longer and increasingly complex periods of development of the growing life form. Another advantage of such compartmentalization was protection of the process of offspring growth, as in the relatively stabilized environment of eggs and seeds.[23]

Regarding the timing of the evolution of information-rich self-replicators

---

[22] Strasdeit (2010): 113.
[23] Sterelny (2011).

encapsulated in protective compartments, a current consensus would suggest a date of at least 3.5 billion years ago.

Transition 2: Unlinked Replicators to Chromosomes

The timing and circumstances of the aggregation of self-replicating molecules, as sketched out above, into larger structures – known largely today as chromosomes, or tightly-coiled packages of DNA – is also of great interest in biology, and it is also poorly understood. As usual, evolutionary explanations for such a transition seek its adaptive advantage, or 'survival value'. In this way, Spanish microbiologist Mauro Santos has suggested that the advantages of linking independent segments of DNA into aggregate structures included the compartmentalization of specific gene functions into distinctive segments of DNA, called 'genes', which increases fidelity of reproduction in the offspring because offspring inherit the entire gene every time, rather than occasionally inheriting only some of the useful DNA elements.[24]

Whatever the advantages, the linking of previously-unlinked DNA fragments into chromosomes was a fundamental process in early evolution, leading to higher-fidelity replication of the parental generation into offspring. Like the origin of groups of self-replicating molecules, chromosomes are likely to have evolved by 3.5 billion years ago, judging by the high fidelity of replication visible in microfossils of that time period.

Transition 3: RNA to DNA

Early life forms, and many that are still with us today (including many viruses), were characterized not by DNA as the molecule of replication, but RNA (ribonucleic acid as opposed to deoxyribonucleic acid). Rather than two phosphate 'rails', as in DNA, RNA is normally single-stranded and has slightly different chemical composition than DNA; however, it is also a self-replicating molecule, and, being simpler than DNA, it is widely considered to be the precursor to DNA. How DNA evolved from RNA is unknown, but not unknowable: recently Italian biologist Ernesto Di Mauro and colleagues found that in certain easily-attained conditions – water temperature under 70C and an acidic environment – RNA molecules overcame a particular chemical reaction barrier and self-replicated not into single chains, but double-stranded, DNA-like chains.[25]

Like the origins of Transformations 1 and 2, the evolution of RNA into DNA is likely to have occurred around 3.5 billion years ago.

---

[24]   Santos (1998).

[25]   See "New understanding of RNA and evolution", December 18, 2008 at http://www.news-medical.net/news/2008/12/18/44398.aspx.

Transition 4: Prokaryotes to Eukaryotes

The fourth major transition included the appearance of the *eukaryotes* in the domain of Earth life. Eukaryotes differ from prokaryotes in many important ways, summarized in Table 2.2 (which is adapted from Table 11.1 of Lenton and Watson (2011): 213).

While it was once thought that the eukaryotes had a separate origin than the much-simpler prokaryotes in the evolution of Earth life, it is now largely believed that they are derived from prokaryotic ancestors. For some, early prokaryotes absorbed some other prokaryotes, which then symbiotically developed, over time, into organelles and other internal subdivisions within a given cell. This *endosymbiotic* theory is most identified with rogue (but respected) microbiologist Lynn Margulis, whose explanation, while contentious, has been gaining ground in recent years.[26] Another theoretical model characterizes the transition as an "information revolution", in which the eukaryote's much richer DNA content is the most important element distinguishing it from the prokaryotes.[27]

Three main lines of evidence are used to investigate the origins and early evolution of the eukaryotes: (a) the *molecular clock* and developmental genomics, which tell us about the ages of certain adaptations and their commonality or rarity in life forms today; (b) molecular fossils, such as *sterols*, distinctive chemical tracers of ancient metabolism; and (c) microscopic fossils.[28] Both the molecular clock and developmental genomics data have identified that prokaryotes and eukaryotes were present by just after two billion years ago, and microscopic fossils of "near-modern" eukaryotes are found not long after 1.4 billion years ago. Somewhere around two billion years ago, then, and without question by 1.4 billion years ago, early eukaryotes – ancestors of species that would one day build spacecraft to take them off of the surface of the Earth – were present on our planet.

Transition 5: Asexual Replication to Sexual Replication

Sexual replication is the shuffling of the parental DNA before the production of offspring. Specifically, it involves two kinds of parents, a male and a female, each of whom independently shuffles their own DNA (which they inherited from *their* parents) before these individuals generate their sex cells, the male sperm and the female egg. Neither male sperm nor female egg can go on to develop into a mature individual, rather they each carry only half the DNA required to build that individual. When the male sperm fuses with the female egg, the two half-complements of DNA are united and the development of the offspring (called a *zygote* at this early stage) begins. All of this is very different from the relatively

---

[26] Margulis and Sagan (2002).
[27] Lenton and Watson (2011): 89–93.
[28] Lenton and Watson (2011): 229–239.

Table 2.2. Chief Distinctions between Prokaryotes and Eukaryotes.

Property	Prokaryotes	Eukaryotes
Cell size	Small: 1–10 microns (human hair width =about 50 microns)	Large: often 10–100 microns or much larger
Nucleus	Absent	Present (nucleus houses nuclear DNA)
Endoplasmic membrane	Absent	Present; also, endomembrane system subdivides internal space of cell
DNA	Single, circular DNA molecule (chromosome) yields relatively small genome per individual	Multiple, linear DNA molecules (chromosomes) contained within cell nucleus yield a large genome per individual
Cytoskeleton (semirigid structure)	Absent	Present
Organelles	Absent	Present (e.g. mitochondria, which manage energy)
Metabolism	Diverse (includes photosynthesizers, autotrophs [CO_2 consumers], and lithotrophs [inorganic molecule consumers])	Mostly aerobic (oxygen-breathing)
Cell wall	Protein	Cellulose or chitin in some organisms
Generator(s) of variation	Mutation & Horizontal Gene Transfer and other environmental factors	Recombination
Cell division & replication	Binary fission (mitosis), producing near-clones	Meiosis (DNA shuffling) producing variable offspring
Flagella	Rotating	Undulating cilia
Respiration	Via membranes	Via mitochondria
Environmental tolerance	Wide range of temperatures, pressures and chemical tolerances, as well as resistance to desiccation (drying)	Rather narrow range of environmental tolerances

simple 'budding off' process of asexual replication, specifically because asexually-reproducing species do not shuffle their DNA: this means that there is normally much less variation in asexual than sexual offspring. This does not leave asexually-replicating life forms without the genetic variation to survive in changing selective environments, however: that variation is achieved by horizontal (also known as lateral) gene transfer, the picking up of free-floating DNA by one kind of prokaryote from another, different kind (we discuss the significance of this phenomenon later). Sexual replication, then, is the origin of sex differences among individual organisms.

Another prominent model suggests that sexual replication confers advantages in parasite resistance as a result of the diversity it generates more readily than in asexual budding.[29] Which of these (and other) models is most correct (and all might be partially correct), at the least we can say that the early evolution of sex is a fundamental transformation in the evolution of Earth life, and it is at least partially a transformation in the nature and replication of information.

Since sexual replication is a common property of the eukaryotes, some suggest that for both theoretical and practical purposes, the evolution of sexual replication and the evolution of eukaryotes themselves should simply be considered together. Whatever the case, we have good data now to indicate that some Earth life forms were replicating not just by simple 'budding off' of near-clones, but also by shuffling the DNA of the parent generation before the production of offspring, some time after 1.5 billion years ago.

Transition 6: Protists to Multicellular Organisms

The differentiation of cells in a single body, and the proliferation of such cells composing such a body, are important elements in this transition, which takes place over 500 million years ago and is best exemplified in fossils of both plants and animals, each of which are composed of many cells of different kinds (human bodies are composed of over 200 kinds of cells). Why single-celled organisms aggregated in the first place is a good question, and several main models have attempted to explain the advantages of such arrangements:

- Grouping allows movement that individuals cannot achieve, as when slime mold cells aggregate into a mass that can push upward through soil, a task impossible for the individual.
- Grouping allows differentiation of function or action by member cell type, resulting in group versatility compared to the individual.
- Grouping allows growth to larger size, allowing capture and digestion of more food sources.

When considering such lists, we must recall that life characteristics such as

---

[29]   Hamilton, Axelrod and Tanese (1990).

multicellularity might have arisen for reasons we have not yet appreciated, and that for every evolutionary advantage there is normally a disadvantage as well. Whatever the case, a recent review of these options suggests that:

> ".. advantages in feeding and in dispersion [in the early evolution of multicellular organisms] are common. The capacity for signaling between cells accompanies the evolution of multicellularity with cell differentiation."[30]

But we should be careful about suggesting that complex interactions only occur among multicellular life forms. A 2007 review reveals that:

> "Our understanding of the social lives of microbes has been revolutionized over the past 20 years. It used to be assumed that bacteria and other microorganisms lived relatively independent unicellular lives, without the cooperative behaviors that have provoked so much interest in mammals, birds, and insects. However, a rapidly expanding body of research has completely overturned this idea, showing that microbes indulge in a variety of social behaviors involving complex systems of cooperation, communication, and synchronization."[31]

Concomitant with cellular aggregation into larger life forms, such as plant and animal, is the differentiation of function or action of cells composing the larger life form. In this way, mammals, for example, are composed of muscle cells, brain cells, liver cells, lung cells, and so on. At some time in animal evolution, these were different cells – rather than tissues *per se* – but they eventually became very distinctive in their functions; and as the 'design' of the life built evolutionary momentum, the various different forms of life composing the whole became so dependent upon one another – in a sort of unintended symbiosis – that they could no longer even exist alone, making multicellularity, perhaps, a given, once certain conditions were in place. Linked with the 'Cambrian Explosion', an adaptive radiation of life forms discussed below, the origins of aggregations of life forms into larger systems called plants and, certainly, animals, is dated to some time after about 500 million years ago.

Transition 7: Eusocial Societies

The origins of sociality in life – life in which individuals have significant interactions with others of their kind – are best known from the fossil record after about 500 million years ago, after the 'Big Bang' of the biological world, the 'Cambrian Explosion' in which Earth life proliferated and became more complex.

---

[30]   Kaiser (2001): 103.
[31]   West (2007): 53.

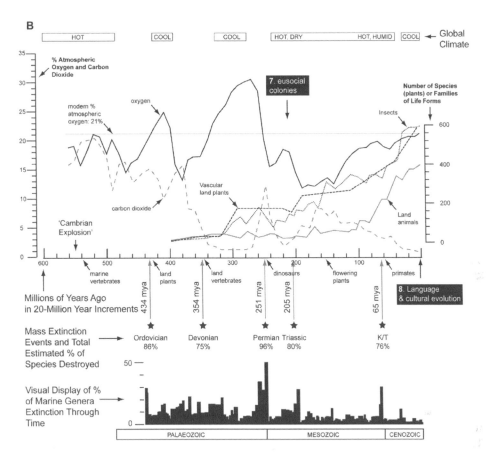

Figure 2.5. History of Earth Life, Part II. The image, adapted from Lenton and Watson (2011) and other sources, focuses on the last 600 million years, and includes evolutionary transitions 7 and 8, changes in atmospheric gasses, the increase of various life forms through time, and some of major extinction events, discussed in chapter 4. Image by Cameron M. Smith.

Specifically, the Cambrian Explosion refers to a period of intense evolutionary radiation of previous forms into new ecological niches and their attendant anatomical and behavioral adaptations. Before the Explosion, life largely consisted of rather solitary, cellular life, or some colonial life forms including the 'Ediacaran biota', plant-like life marine life forms. After the explosion, dated to around 530 million years ago, there appear very different kinds of life, including the *bilaterian* ancestors of all modern animals, characterized by differentiation of the body's tissues and structures such that they possess a front and a back (anterior and posterior), an upper and lower surface (dorsum and ventrum) and a mouth leading to a digestive tract. This is in contrast to earlier life (much of which remains with us today, however, such as the venerable jellyfish), possessing no discernible nervous system or digestive or circulatory organs.

Among the earliest known bilaterians are members of the genus *Kimberella*, a sort of marine snail, covered with dorsal shell plates, that grazed the sea floor over 558 million years ago.[32] The presence of the shelly dorsal protection in *Kimberella* is vitally important. While many explanations of the profusion of life forms after the Cambrian Explosion have been forwarded, one of the most convincing is that it is related to the evolution of ecosystems and predator–prey relationships. Specifically, palaeontologist K.J. Niklas has convinced many in his field that simply as a function of time and variation in the replication of life forms, by the immediate pre-Cambrian explosion times the *morphospace* (possible range of life form shapes) had been filled, such that competition between previously-isolated forms of life began. The runaway effect of such competition included, for Niklas, the development of predation of one life form upon the other and the evolution of predator strategies and prey defenses.[33] If there are inevitabilities in the evolutionary process, they are of this kind, which is a result not of some internal guidance in a centralized, evolutionary scheme, but better characterized as emergent properties of complex systems.

However the Cambrian explosion originated, it certainly established the template for the evolution of most later life forms, including human beings. This is indicated not only in the fossil record, but also increasingly with molecular evidence that even allows us to track the evolution of certain very ancient and widely-shared genes across the entire Animal Kingdom. Palaeontologist Charles R. Marshall has recently summarized the state of affairs:

> "Spectacularly, many key developmental genes and gene families are shared between all animals. Comparative developmental genetics has shown us that while morphologically [in terms of shape] the animal phyla [major groups] might be 'apples and oranges', they are fundamentally comparable."[34]

This allows us even to reconstruct ancient life forms not just with fossil evidence but also with molecular data. Figure 2.6 is a reconstruction, by Cameron M. Smith, of a primitive animal life form of the Cambrian times, based on work by evolutionary biologist Sean Carroll and his colleagues. This incredible image reconstructs an ancient life form from the genes we know it possessed; only a few are mentioned here. We know the animal was segmented from front to back because the *engrailed* gene is so ancient, and we know it possessed a circulatory pump (a heart, which implies vessels to carry a kind of blood) because the *tinman/NK2.5* gene is so old, as is the *ems* gene, related to the production of a nervous system, shown here by a dashed "nerve fiber." And there

---

[32] Fedonkin and Waggoner (1997).
[33] Niklas (1994).
[34] Marshall (2006): 366–367.

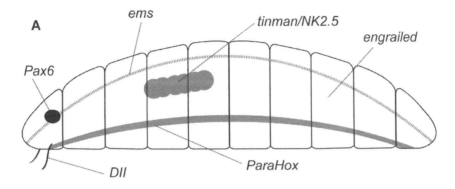

Figure 2.6. Reconstruction of Early Animal Life Based on DNA Data, and Reconstruction of Fossil Primate Based on Fossil Evidence. Top image by Cameron M. Smith, after a reconstruction by Professor Sean B. Carroll of the University of Wisconsin. Lower image by Cameron M. Smith.

is a gut, indicated by the ancient *ParaHox* genes, and body outgrowths, such as feelers, indicated by the *Dll* genes conserved in most life forms today that have such outgrowths. Finally, a photosensitive eyespot is indicated by the venerable *pax6 gene*. Looking into this image we are looking into early animal life as revealed by genetic evidence available only in the last ten years. Below this genetically-reconstructed life form is a representation of an early member of the Primate Order, discussed later in this chapter.

Transition 8: Language and Symbolism

For Maynard-Smith and Szathmary, the evolution of nervous systems led to greater communication among life forms, leading, eventually, to language and symbolism. While this appears generally correct, we feel that anthropology has had a better track record of understanding the evolution of language – which has only appeared in *Homo* – and so we return to this transformation, and its staggering consequences, later in this chapter.

Larger Revolutions in the Evolution of Earth Life

Taking a larger view of the fundamental transformations of evolution, Tim Lenton and Andrew Watson have recently suggested that the transformations mentioned above, while of interest, can be subsumed into three major and distinctive larger 'revolutions' in the history of Earth life, reviewed below.

The first of Lenton and Watson's revolutions is the evolution of *oxygenic photosynthesis*, a process in which light energy, captured by a 'Light Harvesting Complex' – composed of a microscopic protein network of 'antennae' that excite pigment molecules with photons – is transferred to a reaction center where water is oxidized to gas and the carbon of carbon dioxide is reduced to various sugars. This process not only provides energy for the organism, but also liberates oxygen, and for Lenton and Watson this is critical because over geological time that oxygen significantly changed the Earth's atmosphere, resulting in a 'Great Oxidation' that, among other things, led to the evolution of eukaryotes from prokaryotes, discussed earlier. Evidence for the evolution of oxygenic photosynthesis is first unequivocal shortly after 3 billion years ago, in various traces of oxygen as a significant gas in the Earth's atmosphere.[35]

The next great change, the *recycling* revolution, involved the capture of carbon, hydrogen, nitrogen, oxygen, phosphorus and sulphur from the environment as sources of life energy. This was a revolution in metabolism that resulted in a proliferation of life forms, as more food sources became available through new metabolic pathways. Finally, the *information* revolution involved

---

[35]   See Lenton and Watson (2011): 165–179.

the evolution of DNA, a high-fidelity self-replicator, from RNA, leading to greater diversity in life forms.

Lessons from the Evolution of Earth Life

Wherever Earth life originated, on our planet or away from it, it is clear that evolution itself has changed through time, reminding us that the Universe, and even some of its processes, are characterized by change. Early on, for example, fewer energy sources were exploited than later on, and replication of molecules altered from a kind of chemical evolution originating in autocatalytic reactions to a higher-fidelity self-replication process. Also, replication began as asexual budding, then included sexual recombination, and today includes both of these as well as horizontal gene transfer, the acquisition of genetic material even across traditional species lines, during the life course, that are passed on to offspring.

The chief lesson here is that, despite the emergence of some patterning purely as a consequence of how life systems work, there was nothing inevitable about Earth life, and that it evolved in multiple contingent situations over vast periods, as seen in Figures 2.4 and 2.5. Had Earth been impacted by some large body, anywhere in this process, Earth life might easily have been extinguished, and it is reasonable to propose that the evolution of extraterrestrial life might have been terminated by such an event elsewhere in our Solar System or, indeed, the Universe. While life might be probable in particular chemical environments, it does not seem to be inevitable, and extinction events show us that it is certainly not indestructible.

The triumphal "March of Life" that we see in hindsight, assembling our evidence for differences in life over time, 'leading' to humanity, is a self-centered illusion. This highlights the fragility of life and the imperative, we argue, of preserving it, at the least out of self-interest, because it is these chemical processes that resulted, without conscious direction, in our own species. And preserving life in our Solar System, as we will argue later in this book, demands spreading it wide in the Solar System and even beyond it. Only that act can preserve Earth-type life in the long run, considering the fate of the Sun, scheduled to burn out in some billions of years, and in the short run, when we as a species might well meet challenges to survival if we remain constrained to a single planet. Certainly, civilization (as we shall see later in this book) is particularly fragile and liable to disintegration, and civilization itself would be required to establish off-Earth populations of our species.

How can we accomplish the feat of preserving humanity by spreading it from Earth? By adapting to non-Earth environments. To do this, we must understand how humanity itself has evolved through time, and how it adapts; and this requires abandoning common conceptions, such as the dangerous concept of 'destiny', which allows us to sit back and wait for human space colonization to happen as a result of some inborn consequence evolution. Rather, a closer investigation of our evolution – as we shall see – indicates that there was nothing inevitable about it, and that we owe our survival as a genus not to the reactive nature of adaptation seen in most life forms but to *proaction*. This reveals that we humans evolve partly in the ways of other life forms, but that through our evolution our genus, *Homo*, has also found ways around evolutionary factors that have caused the extinction of countless other life forms, ways that shift the adaptive burden from anatomy to complex cultures, behaviors, and technology.

The rest of this chapter reviews our evolution as a genus and species, making it clear that while we are a life form with many commonalities with other life forms – for example, we are built from the same essential DNA molecule as a cactus – we humans need not share the same fate of other life forms. Human evolution, we will see, actively creates options and potentials that are unavailable to other life forms, options and potentials that include the preservation of life through the adaptation to off-Earth environments. Regarding this point, Apollo 15 command module pilot Alfred Worden recently wrote of his thoughts while orbiting the Moon in 1971; note that what he refers to as 'hardwiring' is perhaps valid, but not the same as the 'destiny' concept mentioned above:

> "I turned the cabin lights off. There was no end to the stars. I could see tens, perhaps hundreds of times more stars than on the clearest, darkest night on Earth ... My vision was filled with a blaze of starlight ... Was the space program more than an engineering program – could it be part of our genetic drive? I might be circling the Moon at that moment not because of the politics of the Cold War, but because we are hardwired to explore ... In a few billion years, our own sun will die. Perhaps life wanders from star to star over the millennia, refusing to stay and die? Apollo might be the first step of that hardwired survival instinct."[36]

Early Hominin Origins and Evolution

Humans are one of the over 200 species of the Primate Order, a mammalian division of the Animal Kingdom. We are also *hominins*, large primates that walk upright, and our origins in this group appear in the fossil record some time before six million years ago. Below we review the main adaptations and evolutionary trends of the genus *Homo* – to which all humans belong as members of *Homo sapiens sapiens* – beginning by placing humanity in the biological order of the primates.

---

[36]   Worden and French (2011): 197–198.

Adaptive Evolutionary History of the Primate Order

Primates are mammals characterized by a number of distinctive characteristics, including:

- Emphasis on vision as the main sense.
- Reduction of smell as a significant sense (note the relatively small snouts on primates as compared to, say, dogs).
- Having fingernails rather than claws.
- Diverse diet including grasses, leaves, roots, and other animals.
- Large brains relative to body size, compared to other land mammals.
- Complex social dynamics, including long period of parental care for young.

Primate origins go back as early as 65 million years ago, at the time of the extinction of the dinosaurs. Like many other small mammals of the time, the primates flourished and diversified into many forms, filling ecological niches vacated by the dinosaurs. Based on fossil remains we can reconstruct the early primates, which appeared something like today's lemurs; the lower panel of Figure 2.6 is a reconstruction of an early primate in its *arboreal* (in the trees) habitat. Note that at this early time, the snout is quite large, and the body has not yet made the evolutionary transition to vision as the most significant sense, though the eyes are rather large. Note also the hands and feet, which are very dexterous, facilitating climbing and grasping.

There have been five main *adaptive radiations* of the primate order. Adaptive radiations are significant evolutionary events that result in major changes in the lineage resulting from adaptations to a variety of factors, chief among them being new ecological circumstances. The **first radiation**, around 60 million years ago, established the essential primate body form, and involved dietary changes including an addition of fruits and seeds to the largely-*insectivorous* (insect-eating) diet. The **second radiation**, about 50 million years ago, included the shift of anatomical investment from *olfaction* (sense of smell) towards vision. The **third radiation,** about 40 million years ago, included significant increases in body size – shifting from roughly squirrel-sized to roughly cat-sized – and the evolution of prehensile tails (which are used in climbing, just like an extra hand) among the New World primates, which diverge around this time from the Old World as the Americas are separated by waterways from Africa and Europe. The **fourth radiation**, around 25 million years ago, included the branching of Old World primates into the smaller monkeys that ate a largely leafy diet (*folivory*) and the larger apes, which continued to eat leaves but included perhaps more fruit in their diet. The **fifth radiation**, not long after seven million years ago, included the origin of the hominins, apes that walk upright and are, ultimately, our ancestors.[37]

Figure 2.7 is the essential classification of the living primate Order.

---

[37]   This reconstruction is based on Klein (1989): 95–97.

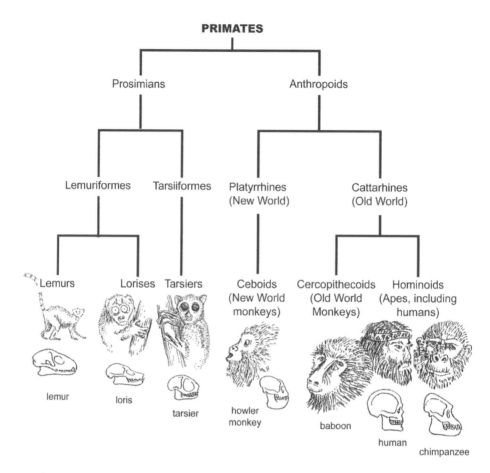

Figure 2.7. Basic Classification of the Living Primates. Image by Cameron M. Smith.

The Early Hominins

Hominins, as we have just seen, originate in the fifth main primate radiation. As recently as the 1920s it was believed that this might have occurred in East Asia, but today ample evidence firmly establishes that this took place in Africa. Specifically, three lines of evidence – anatomical, fossil and DNA – show that around seven million years ago some group of large primates diverged from the rest of the apes. The characteristics of hominins are well-known and relatively easy to identify; they are *bipedalism* (bi referring to *two* and *pedal* referring to the feet, reflecting locomotion on two feet; that is, walking), *dental reduction* including significant reduction in the size of the canines, and *non-honing chewing*, in which teeth are not sharpened when chewing.

Bipedalism is easily identified in fossil remains because it involves significant restructuring of many parts of the body. These include a wide, weight-bearing pelvis because all of the upper body weight is transmitted through the pelvic structures, long, powerful legs, relatively short arms compared to other primates (who use long, powerful arms in locomotion), an s-shaped spinal column with several curves that act as a shock-absorbing mechanism during walking and running, and an arched foot in which the large toe is in line with the rest of the toes rather than *everted*, as in gorillas and chimpanzees, whose large toes stand off to the side of the foot, assisting in climbing.

Non-honing chewing is also evident as the very large canines of other apes – chimpanzees, gorillas and orangutans – are absent among hominins. The canines are there, they are just very reduced in size, as are most of the teeth.

Fossil evidence for the earliest hominins is sparse, but important. The earliest is *Sahelanthropus tchadensis*, known from fossils discovered in Chad, dated to over six million years old. (The genus name, *Sahelanthropus,* refers to the region the material was found [Sahel] and its apelike character in general [anthrop], and the species name, *tchadensis*, refers to the country in which the material was found, Chad). *Sahelanthropus* is currently represented by one largely-complete fossil cranium (though the lower jaw is not present.) While remains of the *locomotor skeleton* (skeletal elements related to bipedalism) have not yet been found, we can be sure this primate was a biped because the *foramen magnum* ('big hole') which admits the spinal cord into the skull is very much under the braincase, indicating an upright posture. We will see, later, that the brains of hominins increase significantly in size over time, but at this early stage the brain of *Sahelanthropus* was about 350cc in size – just under the volume of a soda can, comparable to that of a chimpanzee, and about one quarter the size of a modern human brain.[38]

Another early hominin is known from excavations in the Tungen Hills of West-central Kenya; this is *Orrorin tungenensis* ('orrorin' refers to 'original' in the local Tungen language). This hominin, dated to about six million years ago, was also clearly bipedal – as revealed by the buttressing, overall strength, and fine anatomical details of its *femora* (plural of femur or the upper leg bone).[38]

After five million years ago the fossil record of early hominins improves considerably (Figure 2.8 shows field survey at the important site of Koobi Fora, Kenya). After over a decade of fossil collection in the Afar region of Ethiopia, palaeoanthropologists from a very large, international team recently announced the discovery of *Ardipithecus kadabba* ('Ardi' meaning referring to 'ground' or 'floor' in the Afar language). Dated to about 5.6 million years ago, 'Ardi' was publicly announced in 2009 and quickly became the latest superstar of the world of hominin origins. This is because there are a lot of fossils, from all parts of the bodies of nearly 40 hominins, and while the characteristics clearly indicate some bipedalism, the species probably also spent considerable time in the trees, as

---

[38] Brunet et al. (2002).

Figure 2.8. Hominin Fossil Survey Area at Koobi Fora, Northern Kenya. Photo by Cameron M. Smith.

revealed by its feet, which have a large, divergent big toe as seen in chimpanzees and gorillas.[39]

Why Bipedalism?

One of the great unanswered questions of anthropology is why bipedalism evolved. The question, if framed this way, will probably continue to be unanswered because it is not suited to evolutionary analysis; we might as well ask why hair, or nervous systems, or teeth evolved. It is easy to say these things evolved *to* serve a particular function, like digestion or locomotion, but that is a poor answer; as we have seen, evolution is the *consequence* of replication, variation and selection, and it has no intent or foresight with which *to* create certain structures or drive the evolution of a given life form in any particular direction.

A more evolutionarily-informed way to frame the question of bipedalism is to ask what selective advantages this kind of locomotion provided over others. Such a framing allows empirical tests of the hypothesis. For example, we can note that chimpanzees and gorillas, our closest biological relatives (sharing well over 95%

---

[39] White et al. (2009).

of our DNA) are quadrupeds; they locomote largely on all fours and that locomotion can be quantified by biomechanical analysis. DNA evidence clearly indicates that the hominins diverged from the chimpanzees on the order of 8 million years ago, so we may ask, why did some of our chimpanzee-like ancestors, not long after 8 million years ago, begin to stand and walk, habitually on two feet? What, then, were the selective advantages of this behavior? Modern bonobos (a kind of chimpanzee) often stand and walk on two legs when carrying objects, so we may say that bipedalism 'facilitates' this behavior. Let us add this to a list of proposed advantages of standing and walking bipedally:

- Facilitates carrying items.
- Facilitates seeing over tall vegetation.
- Frees the hands, allowing more complex manipulation of objects.

But in evolution there are few perfect adaptations; it seems that there are costs to all features; in the case of bipedalism, they include:

- Reduction in predator-avoiding speed.
- Reduction in predator-avoiding agility.
- Threatens immobility if even one foot is injured.

Dozens of anthropologists have weighed the various advantages and disadvantages of bipedal locomotion to explain the origins and persistence of bipedalism. For example, Peter Rodman and Henry McHenry have produced biomechanical data suggesting that bipedalism is a particularly efficient mode of walking for mammals the size of early hominins (though it is not particularly efficient for *running* compared to other, non-primate mammals) and that such efficiency would have been selected for in evolution.[40]

Palaeoanthropologist Owen Lovejoy, on the other hand, suggests that with the hands freed to carry things, males were able to *provision* females with food (carried from sites afar) in an exchange for sex, a behavior seen (though without the bipedalism) in many primates today; his model is decades old and influential, and he has recently restated its elements,[41] but it is also – like many theories of early hominin life – difficult to verify in the fossil record. Yet others have suggested that hominins evolved bipedalism by spending a lot of time walking in the shallows of the African lakes, a frankly nonsensical argument called the *Aquatic Ape Hypothesis* (AAT) that has convinced no professional anthropologists, largely because – as today – African lakes in the time of the early hominins supported vast numbers of Nile crocodiles, ambush hunters that explode from muddy water and have no natural predators. Later we will see that while the African savanna was also populated by many predators, hominins devised ways to avoid them.

---

[40]   Rodman and McHenry (1980).
[41]   Lovejoy (2009).

What we can say at present is that no single model or explanation for bipedalism has been widely accepted in anthropology; right now, there is no answer to that question other than the technically-accurate – but somehow underwhelming – statement that it evolved because it provided some (currently unknown) evolutionary advantage(s). We must be careful with such explanations, however; if we look for an advantage in every anatomical or even behavioral characteristic, we might fall into the trap of *hyper-adaptationism*; this is the belief that every evolutionary characteristic exists for a special function, but if we apply that to the human appendix, for example, we see that it is not the case. The human appendix is today an essentially unused and useless vestige of the large intestine; it is not perfectly adapted for anything (although in the past it did serve a digestive function). In the case of bipedalism, however, so many anatomical changes occurred in association with it that we can be sure we are not looking at a vestige or other non-adaptive feature, but a whole complex of traits in a functional whole that reflects a genuine adaptation; locomotion on two feet.

Whatever the selective advantage(s) of bipedalism, they must have been considerable because, as mentioned, a good deal of the skeleton – not just the locomotor skeleton – was evolutionarily restructured over time in the adoption of bipedalism. Remember that this was not an evolutionary plan, and that not even hominins themselves knew this was occurring; all that occurred was that some of our ancestors born with slight variations in their anatomical structures that made bipedalism even a slight bit easier or more efficient than their siblings and other peers, gained some advantage in that way and then passed the genes for such a variation on to the next generation. (Recently a gorilla at a zoo in the UK has been observed to walk on two legs whenever it strikes his fancy; while he is anatomically an habitual quadruped, he is practicing opportunistic bipedalism.) In short, all we can say is that over time, for reasons still unknown, bipedalism shifted from being an opportunistically-used mode of locomotion to the primary mode of locomotion for a number of large African primates some time shortly before about six million years ago.

The Australopithecines

By three million years ago we have excellent fossil material representing another important genus, *Australopithecus*. This genus was named, confusingly, not for being found in Australia, but using the term *austral* to refer to 'South' (Africa), where its fossils were first discovered. However, as we will see, fossils of *Australopithecus* (collectively referring to *australopithecines*) were subsequently found as far north as Ethiopia.

The australopithecines can be divided into two main groups; the *graciles* and the *robusts*. Though the *postcrania* (skeleton below the cranium) of these groups were very similar, the robusts had massive teeth and a fortified skull and face that coped with severe chewing stresses.

Gracile australopithecines

Graciles are well-represented by hundreds of fossil finds. The earliest is *Australopithecus anamensis*, dated to over four million years ago,[42] but much better-known is *Australopithecus afarensis*, dated to somewhat over 3 million years ago named for the Afar hills of Ethiopia where they were discovered in the 1970s. Of the discovery, after weeks of searching the desert terrain, palaeontologist Donald Johanson wrote;

> "The gully in question was just over the crest of the rise where we had been working all morning ... There was virtually no bone in the gully. But as we turned to leave, I noticed something on the ground partway up the slope.
> "That's a bit of a hominin arm," I said ... We knelt to examine it ... "Hominin ... I can't believe it," I said, "I just can't believe it."
> "By God you'd better believe it!" shouted Gray ... his voice went up into a howl. I joined him. In that 110-degree heat we began jumping up and down ... There was a tape recorder in the camp, and a tape of the Beatles song "Lucy in the Sky With Diamonds" went belting out ... At some point ... the new fossil picked up the name of Lucy ... .although its proper name – its acquisition number in the ... collection – is AL288-1."[43]

Over the following years more fossils were recovered by Johanson's team, and *Australopithecus afarensis* is one of the best-known early hominins. The big toe is in line with the rest of the toes, indicating a very modern gait, however the forearms are still relatively long, indicating that Lucy and her kind still spent some time in the trees using powerful arms and hands to grasp and climb. The braincase is a little larger than that of earlier hominins, just over 400cc (now becoming statistically significantly larger than that of our closest living relatives, the chimpanzees), and the body was also slightly larger than in earlier hominins. Though females would have stood just over 1m (about 3.5 feet), males were considerably larger (nearly twice the height of females), a point we will return to later.

Perhaps the most significant aspect of the Lucy finds is that they clearly show obligate bipedalism in the full anatomical commitment or adaptation to this form of locomotion. Lucy an the hominins of her kind walked in an entirely modern way. This can be identified both in the locomotor skeleton and the astounding discovery of fossilized footsteps at Laetoli, Tanzania. Dated to just over 3.6 million years old, the footprints were made by three *A. afarensis* individuals walking across volcanic ash that had just fallen from a nearby

---

[42]   Leakey, Feibel, McDougall and Walker (1995).
[43]   Johanson and Edey (1981): 16–18.

volcano. Notably, the large toes are in line with the rest of the toes, and nearby finds of the remains of over 20 *A. afarensis* individuals confirmed the species as being responsible for the footprints.[44] Other evidence confirming obligate bipedalism in *A. afarensis* are the foot bones and other locomotor structures of a 3.3 million-year-old *A. afarensis* juvenile (estimated to have been about three years old at death) excavated at Dikka, Ethiopia.[45]

Another gracile, *Australopithecus garhi*, dated to 2.5 million years ago, is particularly significant because it is the earliest hominin confirmed to have utilized stone tools. Although stone tools were not found with the australopithecine fossils, the bones of other animals found very close to the australopithecine remains – both horizontally and vertically) – bear distinctive butchery marks, nicks and percussion damage resulting from the use of stone tools to cut muscle tissue from large mammals and break apart their bones to get at marrow.[46] At the nearby site of Gona River, stone tools (but no hominins) were excavated and dated to 2.6 million years ago, currently the earliest widely-accepted date for excavated stone tools.[47] We can say with confidence that stone tools were being used by some kind of hominin, then, by 2.6 million years ago. We will return to the importance of these tools – and what they tell us about early hominin life and adaptation – later in this chapter.

The best-known of the graciles is *A. africanus,* found in large numbers in South African fossil sites such as Sterkfontein ('strong fountain', referring to the natural springs in the area), which has been excavated continuously since the 1930s. Sterkfontein is a wonder, so far having yielded over 9000 stone tools, 500 fossils of early hominins, and thousands of fossils of other animals. The Sterkfontein site is the remains of a complex of ancient limestone caves, where hominin bones were deposited sometimes as hominins died while inhabiting the caves, but also sometimes when big cats, such as sabretooth cats, carried the bones into the caves to gnaw on them.[48] At least one hominin skull fragment has two distinctive puncture marks from the canines of a leopard.[49] Fossil material continues to be discovered at Sterkfontein (and other South African cave sites).

Generally speaking, the gracile australopithecines, dating from roughly four million years ago, were much like chimpanzees or gorillas; large-bodied primates with a diverse diet. However, the graciles differed in that their brains expanded over time – slowly, at first – and of course they were bipedal, spending increasing time on the ground, and not long after three million years ago they were using stone tools and, apparently, consuming animal tissues. We will come to the

---

[44]   Leakey and Hay (1979).
[45]   Alemseged et al. (2006).
[46]   Asfaw et al. (1999).
[47]   Semaw (2000).
[48]   Brain (1981) provides a comprehensive review of the site-formation processes at Sterkfontein and other South African cave sites.
[49]   Clarke and Kuman (2000).

significance of these behaviors, and the evolutionary fate of the graciles, after examining the other two main early hominin varieties, robust australopithecines and early members of the genus *Homo*.

Robust australopithecines

Like graciles, robusts were large-bodied, bipedal African primates; but unlike graciles they possessed massive teeth and jaws as well as a distincitive *saggital crest*, a mohawk-like fin of bone atop the skull (aligned with the anatomical *saggital* plane) that served as an anchor point for substantial chewing muscles (note that the robust's teeth are massive *for hominins*, but share the general hominin characteristic of being smaller than the teeth of other primates). The anatomical characteristics of the robusts indicate a biological premium, in this group, on the ability to chew (the first stage of digestion or metabolism in the primates) a tough diet. This biological investment and adaptation is corroborated by examinations of *dental microwear* which reveal that wear patterns on robust teeth most closely match those of modern rhinoceroses, browsers who eat dryland vegetation.[50]

Robust australopithecines are first observed in the fossil record at about 2.5 million years ago, the date of the fossil remains of the 'Black Skull', a spectacular fossil cranium discovered at West Lake Turkana, Kenya. Though this discovery is *edentuous* (the teeth fell out before fossilization), the dental *crypts* (sockets) are large and clearly of the robust type; and the saggital crest and other buttresses that diverted and absorbed chewing stresses are very prominent, leading analysts to even call it 'hyper-robust'.[51]

Other important robust australopithecines include the species *A. robustus*, a variety in Southern Africa, and *A. boisei* (named for excavation benefactor Charles Boise!). This is an East African variety (note that some palaeoanthropologists suggest that the robusts are so different from graciles that they should be removed from the genus *Australopithecus* and given their own genus, *Paranthropus*, but this is a minority view and most palaeoanthropologists agree that all varieties of the robusts share enough characteristics that they can be referred to, at large, as 'robust australopithecines').

Figure 2.9 shows an image of the cranium of specimen KNMER-406, a million-year-old robust discovered at Lake Turkana, Kenya by Richard Leakey, and below it, a reconstruction of this hominin in the savannah ecosystem.

Robusts have not been found in close association with stone tools, however the argument has been made that in South Africa they (and graciles) rather more often used pieces of bone and horn than stone tools to dig into termite mounds.[52] Like graciles, robusts had brains about the size of chimpanzees and gorillas, around 420cc (we will compare hominin brain volumes later in this chapter).

---

[50]   Grine (1981).
[51]   Walker et al. (1986).
[52]   d'Errico and Blackwell (2003).

Figure 2.9. Fossil and Reconstruction of a Robust Australopithecine. Images by Cameron M. Smith.

Generally speaking, we can think of robusts as something like bipedal gorillas, with massive head and jaw architectures reflecting a significant biological investment in hard chewing. They may or may not have used tools, but none have been unequivocally found with their remains, and, as we will show below, several lines of evidence suggest that if they did use tools, that tool use would have been qualitatively and quantitatively different from tool use in the next hominin species, early *Homo*.

Evolution of Early Homo

Just after about 2.5 million years ago we begin to see fossils of yet another variety of early hominin, this one bearing characteristics that anthropologists consider ancestral to modern humanity. Early *Homo* was a large African bipedal primate similar to the australopithecines, but bearing a number of distinctive characteristics that clearly separate it from the australopithecines, robust or gracile, to which *Homo* is compared below:[53]

---

[53]   Schrenk, Kullmer and Bromage (2007).

- Significant increase in brain volume.
- Increase in stature (body size).
- Further dental reduction (teeth even smaller than those of australopithecines).
- Increased tool use.
- Increase of animal tissues in diet.
- Fully-opposable thumb and forefinger, allowing 'precision grip'.

These changes, some behavioral and some anatomical, are intimately linked. As we will see, stone tools (a cultural invention using material culture) were used to increase animal tissue consumption, which fueled the high caloric costs of the ever-growing brain.

There is considerable debate regarding what fossil material represents the earliest-known member of our genus, *Homo*,[54] but most palaeoanthropologists accept that fossils excavated at Koobi Fora (northern Kenya) and assigned the name *Homo rudolfensis* (Lake Turkana, adjacent to the excavation site, was earlier known as Lake Rudolf), is a good early representative. Dating to 2.2 million years ago, the fossil cranium cataloged as KNMER-1470 has a brain volume of 736cc, nearly two times the volume we see in most australopithecines. Other early *Homo* material, some dating to as early as 2.5 million years (according to some authors), reveals the new characteristics noted in the list above:[55]

- Whereas australopithecine *postcanine tooth area* (entire area of the teeth behind the canine, including the molars and premolars) averages 620mm^2, for early *Homo* it is closer to 525mm^2.
- Whereas australopithecine body height averages 1.39m for males and 1.1m for females, early *Homo* body height averages 1.6m for males and 1.3m for females.
- Whereas australopithecine brain volume averages 481cc, early *Homo* brain volume averages 601cc.

In addition, excavations have shown that a number of early *Homo* sites include substantial numbers of stone tools. Early *Homo*-stone tool associations were found at Afar locality 666 in Ethiopia (2.4–2.3 mya)[56] but large numbers of stone tools found with early *Homo* are best known from the Olduvai Gorge in Tanzania, dated to 2.0 million years old. All of the earliest tools assigned to the *Oldowan* tool-making *tradition* or *industry* (Oldowan here referring to Olduvai Gorge, which has the best examples, while the terms 'tradition' and 'industry' refer to the essential characteristics of these early tools). Oldowan tools, a sample of which is seen in Figure 2.10, are simple but very effective; some are blocks of stone with a sharp edge used to chop, some are flakes with sharper edges used to

---

[54] Grine et al. (1996).
[55] These figures are derived from McHenry and Coffing (2000).
[56] Kimbel et al. (1996).

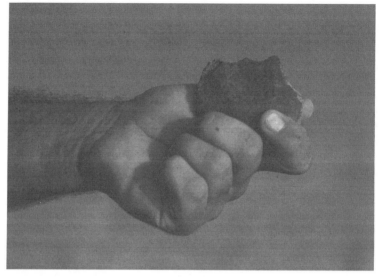

Figure 2.10. Oldowan Stone Tool Excavated at Karari Ridge, Kenya. Photo by Cameron M. Smith.

cut, and others might well have been thrown as weapons. A significant characteristic of Oldowan tools is that they are not symmetrically shaped, a development we will see later.

Sites yielding fossils of early *Homo* often also yield scatters of fossilized non-hominin animal bones that bear cut marks and percussion marks indicating the dismemberment of animals and breaking open of the bones to access marrow; this indicates an increasing reliance on animal tissues in the diet of early *Homo*, that is, after about 2.5 million years ago.

Before going on to examine the world of early Homo in detail, we must examine the evolutionary fate of the australopithecines.

Evolutionary Fate of the Australopithecines

The robust australopithecine is known only from 2.5mya to 1.0mya; after a million years ago, no further robust fossils are known. Nor are robust australopithecine characteristics seen in any species after about a million years ago, and it appears that the evolutionary fate of the robusts is a classic case of extinction; if it evolved into something else, we would know it because its characteristics would be evident in fossils of that 'something else'. Rather, the last robust is known from a mandible discovered at Peninj, Tanzania, dated to just over 1mya. Precisely why the robusts became extinct is stool unknown, but it may be that its apparent anatomical specialization as a dry-vegetation browser might have been a dead end; biologically, dietary specialization can be advantageous in the short run as it reduces competition with other animals for the same resource, but if conditions change such that the food source is cut off, and the organism is *canalized* or evolutionarily 'fixed', it might not have the variability required to adapt to the changed conditions. As we will see throughout this book, that variability – in humans largely behavioral rather than genetic – is of great adaptive significance. Note that, of course, the robusts would not have known about any of this; they would only have perceived (on an individual scale only, not as a lineage with the consciousness of *Homo*) that, for whatever reason, finding appropriate food and mates became more and more difficult. Eventually, the robusts became extinct.

The gracile australopithecine also disappears from the fossil record, but it is largely accepted that this is because graciles evolved into *Homo*. That is, fossils of late graciles and early *Homo* are so similar that the fossils can reasonably be read to indicate the evolution of graciles into *Homo*. This view is not universal; some workers feel that the robusts and graciles were simply large, terrestrial apes that all became extinct. For the moment, the jury, we might say, leans towards graciles as the ancestor of *Homo*. Another line of evidence might be available soon; ancient DNA: 'A-DNA' studies are increasingly common as techniques are improved and as we discover that even very old fossil material sometimes contains – deep inside the fossil – DNA that was not replaced by minerals during the fossilization process. Svante Pääbo, an A-DNA expert who has analyzed Neanderthal DNA dating to over 40,000 years old, suggests that DNA might be preserved for a million years.[57] We are betting (and hoping) that Pääbo is incorrect, and that australopithecine and early *Homo* DNA will be available before long; that might well solve the case of the origins of *Homo* and the evolutionary fate of the gracile australopithecines. At this writing, in fact, the newly-established 'Malapa Soft Tissue Project' is working to recover sequenceable early hominin DNA.

Figure 2.11 shows a wide consensus regarding the relationship of robusts, graciles and early Homo.

The World of Early Homo

Many fossils of early *Homo* have been discovered, and they vary quite a deal in brain volume, tooth size, and so on, leading researchers to suggest that the adaptive radiation of early *Homo* included many different species (though they are all distinctive enough to be considered a genus – *Homo* – distinct from the genus *Australopithecus*). A *species*, we have seen, is a biological group of life forms that breed among their own kind, but not with other kinds; this is easily-enough observed in living populations, but how can we be sure that fossils of early *Homo* (and other hominins) represent different species? Until ancient DNA (hopefully) solves that question, the answer comes from the field of comparative anatomy. In short, if fossil species show the same magnitude of difference that we see in

---

[57] Pääbo (2004).

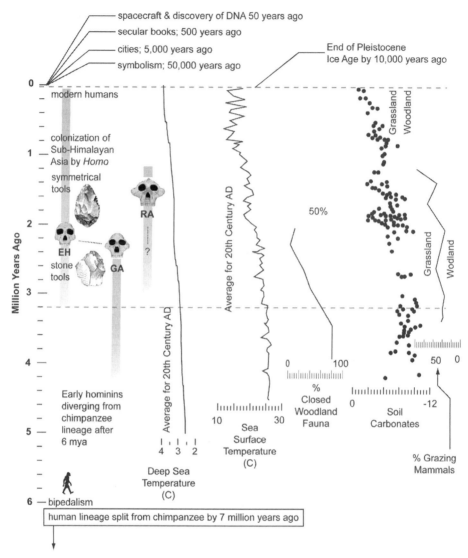

Figure 2.11. Relationships of Hominins Mentioned in the Text, With Information on Tool Use and Environmental Change. Image by Cameron M. Smith with data for deep sea temperature, sea surface temperature, closed woodland fauna, soil carbonates and grazing mammals compiled from various sources, all showing a change from more forested to more open terrain after about 2.5 million years ago.

modern, differing life forms (say, the magnitude of anatomical difference between chimpanzees and gorillas, which puts them in separate genera) then, palaeoanthropologists assign the fossil life forms to different genera (e.g. *Australopithecus, Homo*) or different species (e.g. *Homo habilis, Homo rudolfensis*). While we cannot test these biologically, there is good reason to be cautiously confident in these studies, because their methods have been applied to the classification of many other life forms before DNA analysis was even possible, and now DNA studies to confirm the comparative anatomical studies are, largely, bearing out the differences noted by comparative anatomy.

Most palaeoanthropologists recognize at least two main species of early *Homo*: *H. rudolfensis* (represented by the spectacular KNMER-1470 specimen noted above) and *H. habilis*, which seems to date a bit later than *rudolfensis* and has a slightly larger braincase. For our purposes these (and a few other proposed early *Homo* species) may be grouped and thought of collectively as early *Homo*.

As mentioned, early *Homo* is characterized by a mosaic of anatomical and behavioral characteristics that are connected in that development of one had some effect on the development of others. In this way we may think of the evolution of early *Homo* as a complex, adaptive evolutionary processes composed of many interacting factors. Some of these factors, and how they relate to others, are discussed below, and they are schematically summarized in Table 2.3. (In the table, – indicates that a characteristic is decreasingly significant through time, + indicates that a characteristic is increasingly significant through time and = indicates that some complex interaction of variables results in the indicated change. A final row, indicating the *Extraterrestrial Adaptation*, will be added in *Chapter 8, Distant Lands Unknown*.)

Increased Brain Volume

The average cranial capacity for early *Homo* specimens is debated because, as mentioned, which fossils represent early *Homo* and which represent late gracile australopithecines is debated. Still, most anthropologists would agree that the lower end of cranial capacity in *Homo* begins around 600cc, which is significantly above the australopithecine average.[58]

Note that while brain volume does increase in early *Homo*, the physical anatomy of the brain does not change significantly except for an expansion of the frontal lobes, giving early *Homo* a slightly more prominent forehead than that of the australopithecines. While it would be tremendous if we could read into this details of behavior by identifying 'what the frontal lobes do', it turns out that such simplistic mapping of cognitive (thought-related) capacity and activity with anatomical structure works only in the broadest sense, and in cases of brain injury it has been observed that brain functions once thought to be

---

[58] Dunsworth (2010): 356.

Table 2.3. Changes Associated With Biological and Cultural Adaptations in the Hominins.

Adaptation	Biological Change Involved: Structural/Anatomical	Biological Change Involved: Metabolic/Process	Cultural Change Involved
Preadaptation:	– large canine	– selective pressure for robust body	+ social complexity
Pair-Bonding: by7mya	– sexual dimorphism (size differential between male and female)	+ selective pressure for complex social communications	+ social complexity
Terrestrial Adaptation			
Bipedalism: by 7mya	+ bipedal anatomy – arboreal anatomy	+ thermoregulatory efficiency	+ territorial range + adaptive econiche plasticity
Technological Adaptation	+ finger–thumb opposability	+ hand/eye coordination	+ enculturation time
Tool Use: by 2.5mya	+ brain volume + body stature – overall tooth size	+ caloric needs + caloric needs – digestion requirements (food pre-processed by stone and, later, fire)	+ econiche breadth + econiche breadth + econiche breadth = reliance on technology = reliance on culture use = decoupling of behavior from anatomy
Cognitive Adaptation	+ brain volume	+ caloric needs	+ econiche breadth & or + econiche specificity & or active Niche Construction
Modern behavior: after 100,000BP in *Homo sapiens*.		= more reliance on technology to process foods	
	+ brain volume & brain architecture complexity	= more complex social interactions	= + cognitive variation/complexity = + behavioral variation
	– overall tooth size		= + cultural complexity and enculturation time
	– body robusticity		= + cultural complexity and enculturation time

anatomically localized can in fact be taken over by other regions of the brain. Having said *that*, what does increased brain volume tell us about the life of early *Homo*?

It implies at least two things. First, that the diet of early *Homo* was changing to include more animal tissues. The brain is a calorie-hungry organ that consumes over 20 times the calories of resting muscle tissue, and 25% of our ingested calories are used by the brain alone. Thus, a larger brain meant that early *Homo* would have had to seek maximum calories for minimum effort. When palaeoecologists consider the mixed wooded grassland & thin woodland environment indicated by ancient pollen and faunal remains at most early *Homo* sites,[59] they identify animal fat as the food containing the greatest number of those calories in the entire ecosystem. We saw that some australopithecines used stone tools to break open the bones of animals as early as 2.6mya, but is with early *Homo* that the use of stone tools to butcher animals really picks up, and by 2.0 million years ago it is difficult to imagine early *Homo* surviving at all without stone tools, an issue we will also return to below. For the moment, the lesson is that increased brain volume occurs in tandem with increased use of tools and increased animal tissue consumption, and there are good reasons these three factors influenced one another in a complex feedback relationship.

The second significant implication of increased brain volume in early *Homo* is that it increases the number of neurons available for storage of memory. It is important to remember that the *brain* is an anatomical structure – composed largely of cells called *neurons* – while the *mind* is *what the brain does*. What the mind is capable of doing is to some degree conditioned by the number of neurons available; humans have about 100 billion neurons, while a lab rat has about six thousand times *fewer* neurons, about 15 million. The point is that while we cannot simply equate brain volume with 'intelligence' – a special quality of the mind that we will discuss later – we can say that significant differences in behavior might require significant differences in neuron count.

Increased Stone Tool Use

As mentioned, early *Homo* made and used many stone tools, often by chipping flakes from a core with a hammerstone. Usewear analysis (examination of distinctive wear patterns on stone tool edges resulting in use of those edges for different tasks) has determined that some of these flakes were used for animal butchery and cutting vegetable matter, while some cores were used to work wood.[60, 61] Sharper quartzite tools were likely used more often for cutting while coarse – but less shatterable – basalts were used for heavier tasks, such as

---

[59]   Committee on the Earth System Context for Hominin Evolution (2010).
[60]   Keeley and Tooth (1981).
[61]   Gibbons (2009).

Table 2.4. Non-Hominin Animal Tool Use.

	Insects	Echinoids	Crustaceans	Arachnids	Cephalopods	Gastropods	Fish	Amphibians	Reptiles	Birds	Rodents	Carnivores	Ungulates	Elephants	Cetaceans	Prosimians	New World Monkeys	Old World W Monkeys	Gibbons	Orangutans	Gorillas	Bonobos	Chimpanzees
Drop	x									x				x			x	x	x	x	x	x	x
Propel	x		x	x		x				x	x	x		x	x		x	x	x	x	x	x	x
Shove										x	x	x		x			x	x		x	x	x	x
Wave		x											x	x	x	x	x	x	x	x	x	x	x
Entice	x			x				?		x		x					x	x		x	x	x	x
Beat										x				x			x	x		x	x	x	x
Pound	x									x		x					x	x		x	x	x	x
Lever					x					x		x					x	x		x	x	x	x
Dig				x						x	x		x		x		x	x		x	x	x	x
Pierce										x							x	x		x	x	x	x
Reach										x	x	x		x			x	x	x	x	x	x	x
Probe	x									x				x			x	x	x	x	x	x	x
Abrade										x		x	x	x			x	x	x	x	x	x	x
Cut																	x				x		x
Inhibit		x		x						x	x	x		x			x				x	x	
Bridge									x	x				x	x	x	x	x	x	x	x	x	x
Hang																	x	x	x	x		x	x
Hold	x									x		x					x	x		x	x	x	x
Absorb										x							x	x	x	x	x	x	x
Wipe										x					x		x	x		x	x	x	x
Affix	x	x	x	x	x	x		x	x	x	x	x	x	x		x	x	x		x	x	x	x
Symbol																	x			x	x	x	x

woodworking, indicating that early Homo selected specific raw materials for specific tasks. Many basalt pounding tools are on the order of 10cm in maximum size (about three inches, or the size of a baseball), whereas many quartzite tools are significantly smaller.[62] Table 2.4, which is adapted from Table 7.1 of Shumaker et al. (2011): 215, indicates that while many life forms use tools, it is in *Homo* that we see the fullest use of tools and reliance on those tools for survival.

A significant characteristic of early *Homo* stone tool use is that until about 1.7mya there appear few new forms; the basic chopper, scraper and other

---

[62]　Leakey (1971): 25–37; Blumenschine et al. (2008): 82.

handful of tool types appear to have been the whole Oldowan 'tool kit'.[63] We will see later what happens around 1.7 million years ago.

We can say that *Homo* increased its use of stone tools for a variety of tasks, some certainly related to food processing. Early on, these tools are essentially asymmetrical, and do not suggest a great deal of forward planning, but by 2 million years ago such planning does seem to be indicated by *caches* of raw material – good toolstone – found with early *Homo* at excavations at Olduvai Gorge, Tanzania. Not only this, but early *Homo* carried these rocks several kilometers across the landscape, from the source site (certain outcrops where good toolstone was available) to points on the landscape not in a simple linear fashion, but along paths that would have avoided predation and other risks.[64] All of this suggests a mind – carried by the brain – that is increasing in complexity of action and memory use.

Increased Animal Tissue Consumption

Early *Homo* also began to consume more animal tissue, as revealed by increasing evidence of animal butchery and bone breakage for marrow at archaeological sites dating to after 2.0 million years ago. Many sites at Olduvai Gorge, for example, are complex scatters of stone tools, *debitage* (bits of stone-tool production and resharpening debris), and fossilized animal bone, including much bone that has nicks and other butchery marks, as well as crushed bone indicating marrow acquisition.[65]

The animals from which tissues were removed – fat, marrow, muscle and organs – include land herbivores such as antelope and aquatic herbivores such as hippopotamus. The FxJj50 site at Koobi Fora, Kenya, for example, dated to close to 1.8mya, yielded bones from a variety of savanna grassland animals as well as stone tools and debitage representing use of stones at the site as well as the resharpening of them, perhaps as they were used to dismember animals.[66]

By 1.8mya there is also the fascinating case of fossil specimen KNMER-3733, believed to be a female *Homo ergaster* (one of the several proposed species of early *Homo* that we have subsumed under that more general label in this chapter). The surface of the bones of this partial skeleton did not look quite right to palaeoanthropologist Alan Walker, and in 1982 he and Richard Leakey published the conclusion that the bone lesions indicated *hypervitaminosis A*, a debilitating and ultimately fatal condition resulting from overdoses of vitamin A. Considering where vitamin A is concentrated in the African savannah/woodland

---

[63] Stout et al. (2010).

[64] Blumenschine et al. (2008).

[65] While the site-formation processes are these sites are sometimes complex, and many are palimpsests of many episodes of behavior by either hominins or carnivores, some are genuine traces of ancient butchery by hominins: see Potts (1988).

[66] Toth (1997).

environments that were home to early *Homo*, the authors pointed out that it is found in high concentrations in the livers of carnivores; something that hunters know today, preventing them from eating carnivore livers. In this case, it appears that this female early *Homo* gorged on carnivore liver, a direct indication of hominin animal tissue consumption at nearly 2 million years ago.[67]

While it is clear that animal tissue consumption increased, we must keep in mind three important *provisos*.

First, increased animal tissue consumption does not mean that other foods were not consumed; plant foods, such as fruits, nuts, seeds and tubers (which modern African hunter-gatherers dig from the ground to eat but also for their water content) would all have been important parts of the hominin diet; but so was animal tissue.

Second, we must consider how early *Homo*, significantly smaller than modern humans and with none of our high-tech tools, got their hands on animal tissue in the first place. Certainly early *Homo*, though bipedal, due to differences in maximum speed during running, could not simply chase down, say, an antelope or a zebra; and there is no evidence of projectile weapons – such as spears – at this time. It is increasingly considered likely that hominins began not as hunters, but as scavengers; not only that, but *confrontational* scavengers, like today's spotted hyaenas, who are often successful at driving big cats away from their kills, allowing hyaenas to consume the kill without doing the hard work of the chase. We will come back to this point below.

Finally, keep in mind that while calories are important, even more important is water. Every day, hominins had to consume at the very least three liters – about a gallon – of clean water. This requirement, as well as decisions about stone tool sources and caches (mentioned in the previous section), predator distribution on the landscape (most big cats rest under vegetation during the day), and the distribution of fresh kills on the landscape (perhaps made visible by circling carrion birds) would all have entered into hominin decisions regarding their daily lives, as well as, presumably, memory archived in the physical brain. Considering all of these factors, we can see that early *Homo* did not evolve 'on' an African savannah environment, like an actor on a stage, but *in* such environments, as members of evolving plant–animal communities.

Increased Body Stature

With the beginning of *Homo* we also see an increase in overall body size, often referred to as stature. Palaeoanthropologist Henry McHenry has reconstructed average male height at 1.6m (5.2 feet), with females averaging 1.2m (3.9 feet),[68] and other researchers have found that this stature is found largely in the

---

[67]   Walker, Zimmerman and Leakey (1982).
[68]   These figures are from McHenry (1991).

elongation of the legs (as opposed to, say, the spine); at least one early Homo fossil tibia (lower leg bone) is of a size indistinguishable from that of modern human populations.[69] Though early *Homo* was considerably shorter than many modern humans, it was, as a genus, considerably larger than any australopithecine.

The main implication of this increase is that these larger bodies required more calories to fuel; a modern human body requires about 1000 calories daily just lying and resting, and can demand 5000 calories a day (or much more) if doing strenuous work. Early Homo had to obtain these calories – and fresh water – every day. As mentioned, archaeological sites revealing early *Homo* using stone tools to butcher large animals suggests that, increasingly, a significant number of those calories were derived from animal tissues obtained not – at least at first – by carnivore-like predation, but by scavenging the remains of big-carnivore kills as well as (increasingly) confrontational scavenging: driving big carnivores from their kills. It has been pointed out that such activity would have favored larger, more intimidating (and capable) bodies over time,[70] and again this characteristic – increased body stature – is not an isolated occurrence but must be seen as one of many interacting factors in the evolution of early *Homo*. The larger body was selected for, presumably, because it made early *Homo* more effective at confrontational scavenging, which was needed to gain the calories for the burgeoning demands of the larger brain and larger body – a feedback cycle.

Decreased Tooth Size

The well-documented decreased tooth size in early *Homo*[71] – as compared to australopithecine teeth – is argued to reflect decreasing biological investment in tooth structures as the means of survival in early *Homo*. Though the teeth were still a little larger than those of modern humans, it is clear that the teeth of early *Homo* were not as primary an investment for survival as, say, the teeth of carnivores, whose large canines are essential for quickly killing prey and whose blade-like post-canine teeth are needed to slash meat into chunks that can be swallowed. Increasing stone tool use over time, many anthropologists argue, is responsible for the continual decrease in tooth size in the genus Homo over time; rather than using our teeth to process our food, for example, tools – which will eventually include fire and cooking – are used to process food before it even enters the digestive system.

Here again we have a case where an anatomical characteristic (tooth size) is affected and intimately connected with a behavioral and – ultimately – cultural characteristic, that of making and using tools to survive.

---

[69]   Haeusler and McHenry (2004).
[70]   Brantingham (1998).
[71]   Leakey, Tobias and Napier (1964): 9.

Early Hominin Ecology, Behavior and Culture

The complex mosaic of anatomical, behavioral and cultural changes we see in the fossils and archaeological sites of early *Homo* speak clearly in one regard; this is a substantially new creature on the African savannah, in biology and behavior. On the other hand, it is difficult to identify just how certain variables affected others, in a complex system of feedbacks.

To understand this system it is significant to envision, as we have mentioned, early *Homo* evolving not *on* the savannah, but *in* the savanna, as one element in a complex ecosystem. No life form known on Earth exists in complete isolation; even insects living deep inside caves carry intestinal microbes with which they *coevolve*. Thinking in terms of coevolution considerably helps to comprehend the evolution of early *Homo*.

A wide variety of faunal and floral data have allowed palaeoecologists to identify that beginning around 4 million years ago (when there were plenty of large, bipedal primates in Africa, as we have seen), significant ecosystem changes were occurring on the continent; these included the fragmentation of many heavy forests and their replacement by more open landscapes dotted with trees, cut here and there by rivers that meandered down from highlands and fed large lakes. This savannah ecosystem was colonized not just by hominins but a wide variety of herbivores subsisting on vegetation, and carnivores, subsisting on the herbivores.

How did hominins fit in? Every species in an ecosystem has an *ecological niche*, a specialty. For the robust australopithecines, the niche is revealed in their massive, grinding dentition, reflecting a dry food diet, perhaps most available in the open grasslands. For the graciles and, later, early *Homo*, the dentition shows less such specialization and, instead, increasing reliance on tools (used to procure calories from animal tissues) to survive. Palaeontologist R. Dale Guthrie has recently pointed out that the herbivores that colonized the savanna all had formidable biological defenses, such as horns, that gave them protection from the many predators of the savanna; these herbivores also had special digestive anatomy that allowed them to consume grasses and leaves, something probably unavailable to early hominins. Therefore the savanna herbivores had biological defenses that helped them avoid becoming a meal, and internal anatomy that allowed them to digest a meal. How did early hominins do the same? Consider the carnivores they had to cope with on a daily basis:

> "The predatory techniques of these [African savannah] carnivores varied considerably. Canids and hyaenids became *coursers*, constantly testing the herds for the halt and lame, and trying to catch individuals at some disadvantage that would allow predatory strengths of endurance and stamina to prevail in long pursuits. Large felids relied on *stealth* and special anatomy that allowed a drag-racer style of acceleration. The latter was bought with a loss of endurance, so felids had to start a chase at quite close quarters. To attain that proximity

they evolved exquisite patience and stealth. Some, like hyaenids and canids, were scent *trailers*; others, like cats, developed penetrating nighttime vision."[72]

Guthrie suggests that hominins used sharp thorn bushes to prevent predator attack; lions, he notes, do not pursue antelope into thorn bushes and do not molest livestock enclosed overnight in a *kraal*, the traditional African livestock shelter made of thorny brush. He goes on to suggest that, lacking the biological defenses of every herbivore on the savanna, hominins must have used tools to survive. Whatever tool was used, it seems clear that only such tool use could prevent predator attack. Not becoming a meal, then, was dealt with by hominins by tool use. Defensive tools, Guthrie argues, could include thorn-bushes that modern predators actively avoid; offensive tools could have included thrown stones.

In the same way, hominins lacked the killing dentition, speed, night vision, agility and other characteristics of the top carnivores – think lions and sabertooth cats – that shared the savanna. But to fuel their ever-larger brains and bodies, hominins needed the same animal tissue as those large carnivores, putting small bipedal primates into direct competition with top predators. Again, tool use was probably the key to survival; the concerted use of stones (thrown), pikes (long, sharpened sticks), and dense thorn bushes, Guthrie and others argue, early *Homo* would have had a decent chance of driving big cats and other carnivores from their kills (something that modern humans still do in Africa, armed only with spears, and not firearms), particularly if they confronted young carnivores, or females who are less tolerant to confrontation than males. Getting a meal, then, was also dealt with by hominins by tool use.

Many excavations reveal that over time, stone tool use became more significant in the life of early *Homo* and this suggests to many that the culture of early *Homo* must have become more elaborate through time. Recall that culture is – among other things, as we saw in Chapter 1 – the set of information used to guide behavior, and that it is stored physically in the brain (at least until external memory storage, such as cave paintings, were invented, which is millions of years after the time of the earliest hominins). As hominin stone tools became more elaborate, then – symmetrical tools appear by 1.7mya – the cultural instructions for survival must also have become more elaborate. Culture as information in the brain was growing, and there would have been a premium on passing that cultural information on to the offspring; this is done not with genes – again, as we saw in Chapter 1, cultural information does not ride the genes – it is done through social communication. For most primates, that communication is bodily, gestural, and includes some vocalizations, but only in *Homo* is there the distinctly complex system of communication called *language*. We are nearing a full discussion of the evolution of language; for the moment, it is sufficient to say that the origins of language may be in the increasing complexity of culture

---

[72] Guthrie (2007): 138.

Figure 2.12. Fossil and Reconstruction of an Early Member of the Genus *Homo*. **Images by Cameron M. Smith.**

associated with hominin colonization – and all of the defensive, offensive and food-getting behavior required to do this – of the expanding grasslands of Africa, some time after about 4 million years ago. Below we will see the long-term result of this adaptive suite, wherein the genus *Homo* expands out of Africa, initiating a range expansion that eventually includes the entire globe and, we argue, must include environments off of Earth.

Figure 2.12 shows the spectacularly well preserved cranium and mandible of an early member of the genus *Homo,* and reconstructs that proto-human in the African savannah. Interestingly, what was going on in the mind of early *Homo*, what is of real interest because it will be the mind and its inventions that really adapt humanity for survival, rather than our biology, is no longer a mystery wrapped, as it were, in an enigma. In the next chapter we review the most recent attempts to understand the evolution of humanity's adaptive ace-up-the-sleeve, a highly intelligent mind.

Adaptive Lessons for Human Space Colonization

What lessons can we take from this survey of the origins of our genus? First, it is clear that while humanity is a member of the Primate order of mammals, by some time over 2.0 million years ago some group of primates commenced an evolutionary revolution; the intensive use of objects to supplement the anatomy for survival. This, we might say, was the *invention of invention* and it begins our genus' long-term engagement with objects, materials and techniques, a deep reservoir of experience that we draw from even today, as we manipulate even single atoms. Space colonization will, of course, be intensely technology- and materials-oriented, and the various archaeological lessons of millions of years of engagement with objects and techniques should be mined for what they can tell us about human technological adaptations in the future.

Second, we see that through time our genus increasingly shifted the bulk of our adaptive means from biology to behavior, the *decoupling of behavior from anatomy*. This profound shift further differentiated *Homo* from other primates, and its most extreme development is seen in the human exploration of the moon nearly half a century ago, in which bubbles of human-equivalent environments (in terms of pressure, breathing gas, temperature, and so on) were transported off of Earth to another celestial body. This ability to survive where the physical body could not was due to proaction, also rather unique in the animal kingdom. As we shall see also in the next chapter, this proaction has been necessary to human survival in the past, and will be important to cultivate in the project of human space colonization. As we shall see, also, in Chapter 6, ensuring proaction might require special cultural traditions, and we should not take it for granted.

Third, both of these points indicate the critical faculty of complex cognition as the key characteristic of our genus. We investigate the lessons of this cognition in the following chapter.

3 The Adaptive Suite of Genus *Homo:* Cognitive Modernity and Niche Construction

"... 'human nature', viewed in the context of evolution, is marked by its flexibility, malleability, and capacity for change. The fate of the human mind, and thus human nature itself, is interlinked with its changing cultures and technologies. We have evolved into the cognitive chameleons of the universe. We have plastic, highly conscious nervous systems whose capacities allow us to adapt rapidly to the intricate cognitive challenges of our changing cognitive ecology ... the human brain itself has remained unchanged in its basic properties, but has been affected deeply in the way it deploys its resources."

Merlin Donald[1]

In the previous chapter we saw the early evolution of our genus, and began to see some glimmers of behavior that seem familiar: the making and use of stone tools, in particular, reminds us of our dependence, today, on technology to survive. In this chapter we show how the evolution of modern human cognition – characterized by symbolism and language used in concert in *cognitive modernity* – led to the staggering power of yet another adaptive tool, *niche construction*, by which the genus *Homo* colonized the globe despite its small numbers and relatively frail body. This will give us an evolutionary context for the colonization of space, which will be the most thoroughgoing and intensively proactive case of niche construction in the history of our genus.

Evolution of the Modern Mind

Modern investigation of the evolution of the human mind is often identified with *evolutionary psychology*, most prominently identified with Linda Cosmides and John Tooby at University of California Santa Barbara's Center for Evolutionary Psychology. This field examines the human mind as the product of evolution, and the approach has made great progress in unpacking the

---

[1]   Donald (2004):35. Donald's book, referenced in the rest of this chapter, is Donald (1993).

concept of intelligence, and laying out a basic understanding of how we humans think. Many anthropologists today, however, feel that 'evol psych' attributes too many of our current modes of thought to roots in ancient foraging behavior: critics ask how algebra was derived from the core foraging decisions of the early *Homo* era that seem to be the focus of 'hardline' evolutionary psychology. Others suggest that evolutionary psychology does not sufficiently use what we can learn from archaeology, which studies the artifacts that were made by the very minds that evolutionary psychology investigates. While Cosmides and Tooby have relatively recently defended their approaches, cognitive archaeology today, as we will see, is taking a rather different – though still evolutionary – approach.

Despite these provisos, the evolutionary approach does guide most approaches to the mind – what the physical brain *does* – today. It is important to say a few things about what this evolutionary approach to the minds is not: it is not *sociobiology*, the concept that most behavior is genetically hardwired, nor is it *environmental determinism* which lays all behavior in the hands of environmental influences; it is not *unilineal*, suggesting that all minds and varieties of consciousness will necessarily develop in the same way; and it is not *flat evolution*, attempting to assign an evolutionary advantage of every facet of how modern humans think. The mind has been shaped both by the constraints of its evolutionary history and by the profound effects of culture, and this mosaic of thought will not be explained by any single or simple model. Nevertheless, because both the brain and the mind evolved, they must be studied as products of the evolutionary process.

Evolution is characterized by change, so an evolutionary approach to the modern mind begins with the deceptively simple questions:

- What changed in the mind, through time?
- What was the evolution of the modern mind about?

The Donald Model: Representing Reality

Canadian psychologist Merlin Donald's answers, presented in his provocative 1991 book *The Origins of the Modern Mind*, is that the evolution of the modern mind was fundamentally about the ways that the mind *represented* its experiences. His model outlines three major developments in how the mind, over time, managed the information stored in the brain, with attendant new states of consciousness.

Donald begins with the minds of the late australopithecines and earliest *Homo*, both of whom we met in the previous chapter. Based on their limited tool use and the apparent simplicity of 'Oldowan culture' (Figure 2.10 shows a stone tool made by early *Homo*), Donald suggests that mind had important similarities with that of modern chimpanzees, who excel at perceiving events, but do not retain most of those events in long-term memory, nor do they think about events that might occur in the far future: chimps taught to use sign language most commonly use it for direct requests for food, treats, tickling, and so on. Donald calls this *episodic consciousness*, life lived in a sort of bubble of short-term, small-space perception with comparatively limited memory recall.

The first cognitive revolution, Donald suggests, took place some time after 1.7 million years ago, and is first materially evident in the first symmetrical stone tools dating to that time. For Donald, such symmetry indicates a fundamental change in the mind, termed *mimetic representation*; in this mode, *Homo* did not just recall experiences in an automatic, instinctual, and *reactive* way, but *proactively*, by consciously choosing to select past experiences (for example, how to shape symmetrical stone tools) from many memories of past experience, just as today we select a story – of all the stories we know – to make a certain point.

Communicating intentionally-retrieved memories to other *Homo* would have required some kind of *re-presentational* act, commonly thought of as language. But Donald suggests a precursor to language, called *mimesis*: a pre-linguistic communication based largely on gesture, and perhaps some vocalizations. By voluntarily re-presenting past experiences (the act of representation), early *Homo* broke out of the bubble of episodic consciousness inhabited by all other creatures. Now the mind could selectively remember experiences, picking them from memory archives, and then re-present those experiences, to themselves, to others or to both. Because hominins did not evolve in a vacuum, but were highly social creatures, such representational activity would include communication of past events to others in the group, guided by an increasingly rule-bound system called *grammar*. For Donald, mimesis was something like the invention of a crude dictionary for the mind of early *Homo*, a guide to the specific meaning of certain acts and gestures: perhaps some representations could even be strung together as acted-out sequences of symbols much more complex than any seen in the rest of the animal kingdom. Rhythm, Donald suggests, would have been an important organizational tool for mimetic acts of representation.

As mimetic representations became standardized, more complex, and even completely abstracted – the gestures, for example, no longer bearing resemblance to the thing they represented, such as fear, or anger – there arose a need for organization of the clutter of symbols in the mind. The evolutionary solution was lexical invention, the invention of symbols far richer than the rather literal metaphors and one-dimensional references of mimesis. It is fascinating that this occurred not by making the definitions of symbols more concrete, but by making them 'fuzzier'. We can think of lexical invention as the appearance of 'velcro' symbols (which stick to other symbols) in contrast to the 'teflon' (non-sticky) symbols of mimetic consciousness. For Donald, language was certainly significant – it facilitated the communication of these ever-more-complex thoughts – but it arose as a subsystem of mimesis, as a more efficient way to represent increasingly-complex sets of voluntarily-recalled memories. As mimesis broke the mind from the bubble of episodic consciousness, lexical invention broke the mind from the rather literal world of mimetic consciousness. It allowed representations, such as words, to be used in a wide variety of contexts, rather than always being restricted to specific contexts. Lexical invention may be thought of as the invention of a conceptual thesaurus, to accompany the conceptual dictionary invented with mimesis. In a spectacular snowball effect, lexical invention linked ideas to other ideas, forming complex networks of memory and representation that deepened the meaning of any representation, and promoting enormous variation in the ways that each mind thought, as individuals made new connections based on their own unique experiences.

The complexity implied by lexical invention is breathtaking, and once again, the riot of symbols in the mind, now even intertwined with others in an array of contexts, cried out for organization (a better way to say this, in strictly evolutionary terms, is that when a better method of memory organization occurred, it spread quickly, conferring its advantages to all that carried it and 'swamping' less effective methods).

According to Donald, that organization was achieved by the development of *myths*: complex narratives that integrated and organized the enormous bodies of knowledge being constructed by lexical invention. This was the origin of *mythic consciousness*, in which memories were integrated into specific narratives, told and retold as cultural models of what the Universe was like, and what to do about it.

These myths can be thought of as a kind of encyclopedia to accompany the dictionary and thesaurus of the mind. The mimetic dictionary indicated that A meant B; the lexically-invented thesaurus expanded meaning by saying that A could mean B or C or D, depending on circumstances; and the swiftly-following mythic encyclopedia organized A, B, C, and D (and so on) into integrated narratives that gave order and sense to the Universe.

Donald's third transition was another revolution in the way memory was accessed and represented. This was the invention of *external memory storage*, a concept so familiar to us today that we hardly appreciate its significance. Once again, as the mind became crowded with mythic narratives and an enormous body of human knowledge, a system arose to organize that information. This time, however, the solution was technical rather than biological.

Painting mythological narratives on cave walls and cutting notches into tablets of bone – each as a record of some event – had the profound effect of moving memories *outside* the body. Stores of information were no longer limited by what early humans could physically remember, or even distribute among several memory-specialists (primitive bards, we might say) within society. Now, there was effectively no limit to the amount or detail of information that could be stored and recalled at will. And while previously myth simply *had* to be heavily metaphorical – because even a bard can only remember so many details – now infinite detail could be recorded, limited only by the ingenuity of those who invented the external memory devices and systems.

Cave paintings, hieroglyphs, the modern alphabet; for Donald each is an external memory system used to exceed the biological limitations of the human brain, and our refinement and use of them led to what Donald calls the most recent mode of Western thought: *theoretical consciousness*. The transition to this state was a very recent result of one of the unique properties of complex writing:

the decontextualization of information. Whereas myths – transmitted largely by oral tradition, and for thousands of years served only by rather low-fidelity external memory systems, such as cave paintings, or even hieroglyphs – can only be understood in their own cultural context, completely abstracted writing systems allowed the world to be broken down into finer and finer perceptions, perceptions that could be understood – at least among cultures using writing systems – regardless of the cultural context of their authors. Now memories (and other information, such as measures) could be decoupled from mythological contexts, and thought of thought in abstract, theoretical terms.

Such thinking led to a variety of consciousness that today puts a premium on thought-integrative, information-management skills, rather than rote memorization. Intelligence, today, is about the integration of enormous bodies of information, and knowing where to get specific other information, a point we return to later in this chapter.

Echoes of Representation

Donald does not suggest that each new variety of consciousness simply steamrollered the last. Rather, new varieties (cognitive tools) were added to the old varieties. We invoke our episodic consciousness, for example, when intensely engaged in a single task, as when an athlete shuts out all other thought in a burst of effort. But using a modern mind means that mixed into even that intense, singular effort will be thoughts from other, later varieties of consciousness: a knowledge of the social or historical aspects of the game or competition, for example. We also use mimetic skills every day as we communicate with other people with a rich array of non-vocal gestures. Furthermore many sports, manual crafts, dance, and the creation of visual art, Donald points out, each have little or no speech component, but they remain critical in modern life and they are all learned and executed largely by the mimetic skills devised by the evolving mind over a million years ago. Finally, modes of thought developed during our long period of mythic consciousness continue to condition the way we think, particularly in narrative arts – such as literature – where the skills we developed over thousands of years of organizing our models of the Universe with lengthy, metaphor-rich and memorable metaphors continues to serve and delight the modern mind. From personal relationships to international nuclear non-proliferation meetings, what humans do with language is, basically, tell our stories, negotiate their content to an agreeable truth, and proceed with our objectives. To do this skillfully requires tapping into our precious skills of narration, evolved long ago in the telling of idea-integrative myths.

Finally, the theoretical and scientific thought that allows us to contemplate real (rather than mythical) aspects of our Universe – such as the properties of other planets, or the ecology of our own, or even the nature of our own evolution – are a continuation of the re-integration of decontextualized information made available by writing and other externally-stored memory.

In Donald's model, today the modern mind switches from one variety of consciousness to another, as if we are changing TV channels: engagement in exhilarating, intensely episodic activities, such as leaping from a diving board; participation in mimetic metaphorical ritual, such as dance or marching or chanting at a sports event; immersion into, and contemplation of, rich mythic narratives, both fictional and real, as when we read a long piece of fiction; and deep contemplation of entirely theoretical problems, as when we study calculus or physics or consider the distant future of our own species, and such specifics related to it as the colonization of space.

But while the human lineage had first one, and then two, and then three of these modes at a time, we now have all four – episodic, mimetic, mythic, and theoretic – and that, in Donald's stimulating model, makes the modern mind unique. Modern humans constantly till over our experiences, combining new ones with representations of old ones retrieved from all manner of memory stores, both biological and external, to create new worlds of meaning, and layer upon layer of metaphor. Compared to all other animals – all of whom inhabit an essentially impenetrable locally-bounded (small-space and short-time) bubble of consciousness – the hallmark of the modern mind is this constant and extremely productive integration and reintegration, all courtesy of the various, evolved modes of representation of information, of our stored memories, which constitute representations of our cosmos. And we must remember that the larger lesson here is that this adaptation was more cognitive than anatomical. While certain brain architecture was required for modern consciousness, much more important was the activity of the brain, that is, the mind. We know this because parts of the physical brain can be disturbed or damaged, and yet others sometimes take over the functions lost by that injury. In turn, the lesson of *that* is that this evolution of cognition is, essentially, cultural adaptation, rather than strictly biological, something we have seen in plenty of cases so far.

The Mithen Model: Cognitive Fluidity

An alternative to Donald's model – or perhaps complementary with it – is that proposed by British archaeologist Steven Mithen. For Mithen, the structuring variable in the evolution of modern cognition was not how we represent our experiences, but the degree to which different varieties of intelligence communicated with one another. Mithen's theory, laid out in his 1996 book *The Prehistory of the Mind*, is grounded in the well-established observation that modern minds are 'modular', composed of various domains of expertise or intelligence.[2] Many basic domains of intelligence have been named, but Mithen boils them down to four main types: *linguistic* (use and comprehension of language), *social* (managing interpersonal relationships), *technical* (use and

---

[2]   Mithen (1999).

manipulation of objects), and *natural-history* (understanding cause-and-effect relationships in the natural world). The modern mind, Mithen argues, is distinguished from all others by *cognitive fluidity*, the ability of these modules to communicate with one another.

According to Mithen, early *Homo*, like all other large social primates, had a well-developed social intelligence by 6 million years ago. But by 2.5 million years ago *Homo* started relying on stone tools survive, butchering animals scavenged from big-cat kills. This, in Mithen's model, was the enhancement of existing but not previously-indispensable technical and natural-history intelligences. *Technical intelligence* conditions the ability to make and use tools, and this certainly increased as *Homo* increasingly relied on tools, rather than their biology, to adapt to its environments (for example, the use of stone tools rather than sharp teeth, or claws). This was what we phrased *the decoupling of behavior from anatomy*, a potent distillation of the important changes at this time.[3] Natural-history intelligence also increased, being the capacity for observing the world (in particular, plants, animals, and the 'lay of the land'), and understanding how its parts interacted in space and time.

In Mithen's model, then, in the mind of early *Homo* were three of the four modules of modern intelligence, but those intelligences remained isolated from one another. Mithen's metaphor for the mind is that of a cathedral; in the early cathedral, intelligences were compartmentalized as separate, walled-off compartments, each for special purposes of the cathedral-at-large, without doors connecting compartments – without communication between intelligences.

This cognitive isolation lasted for a vast period, throughout that of *Homo erectus*. What is strange about *Homo erectus* – and compared to the modern mind it is very strange indeed – is that while they made sophisticated stone tools (symmetrical hand-axes about the size of an axe blade), they used them for over a million years without ever innovating a new design. Their technical intelligence was far beyond that of any other animal, but there is no sign of the almost continuous innovation found in the cognitively modern mind, innovation we would expect to see reflected in stone tool design, for example, or the invention of tools from other raw materials. Pre-modern minds thought about making a stone tool, for example (technical intelligence), but rather than thinking about the specific animal they would butcher with that tool (natural history intelligence) at the same time, those intelligences – for this vast period of stasis – remained compartmentalized. Archaeologist Clive Gamble of the University of London (Royal Holloway), a world authority on Middle Humans, has described the sum of these early minds as a '15-minute culture', characterized by routinized actions and a striking lack of innovation. Quoting philosopher Daniel Dennett, Mithen characterizes such pre-modern minds mind as possessing "*rolling consciousness with swift memory loss.*"

---

[3]   See Pilbeam (1998): 526.

For modern cognition, Mithen suggests, one must have *cognitive fluidity*. Continuing his cathedral analogy, Mithen likens this fluidity to doors being opened up between compartments in a cathedral (the mind) that was becoming increasingly complex as its functions and activities became more complex. According to Mithen, fluid communication between the modules of intelligence only occurred somewhere in the last 200,000 years, partly as a result of language.

According to Mithen, early language was a sort of pre-modern 'vocal grooming' that complemented physical grooming – such as when one chimpanzee picks through the hair of another, cleaning it of parasites and building a social bond – as hominin social groups became larger and their societies more complex. As this socially-rooted language became increasingly important, particularly in behaviorally modern life,[4] bits of information about things *other* than social grooming began to 'slip' into spoken communication (how this slipping occurs is fascinating, but beyond our scope in this book). Information from the domain of natural-history activities, for example, began to slip into the domain of social-grooming information. The resulting 'cross-referencing' of thought was profound, and it flung open doors to a vast array of entirely new realms of thought. For example, such fluidity allowed not thinking *just* about tool-making (technical intelligence) or *just* about elk (natural-history intelligence), but *about both at the same time* (being cognitively fluid): that could allow you to specifically make certain types of tools for certain types of activities, in this case, tools specifically for dealing with elk. Or imagine thinking about social, technical, and natural history domains *all at the same* time: thinking about people, tools, and lions simultaneously, for example. Only this kind of cognitive fluidity, according to Mithen, can account for the explosion of rich symbolism associated with Modern Humans, like the lion-person figurine found at Holhenstein-Stadel Cave, in Germany, dated to over 33,000 years ago, depicted in Figure 3.1. The advantages of cognitive fluidity were enormous. Customizing tools for certain plants, animals, and tasks, both social and technical – rather than working with the general-purpose Neanderthal tool kit, for example – opened a nebulous array of new social and ecological opportunities.

The growing complexities of myth, ritual, and symbol, materially encoded in artifacts, solidified and deepened bonds between human groups, effectively increasing their foraging territories and efficiency. Imagination and innovation, the results of language, built the worlds and vehicles that have carried *Homo* to where we are today, for better and for worse.

Echoes of Fluidity

In the same way that echoes of early consciousness are heard in the mimesis and mythic narrative we still use today, Mithen suggests that our modern minds also

---

4   Dunbar (1992).

Figure 3.1. Lion-Person Figurine Over 30,000 Years Old. Drawing by Cameron M. Smith.

carry artifacts of ancient consciousness. Humor, he points out, is often an illustration of inappropriate crossing of domains of intelligence. When Don Knotts, for example, playing the bumbling deputy Barney Fife in the *Andy Griffith* television show, cringes at the door of his precious new car being slammed – as if he himself were being hit – we laugh not because the car is being hurt, but because Barney is inappropriately mixing information from the technical domain (the car) and the social domain (the feeling of pain). In another joke, when a kangaroo complains about the cost of a drink in a bar, we laugh not because the drink is overpriced, but because kangaroos (thinking about animals) and bars (thinking about people) are inappropriately mixed. Cognitive fluidity, according to Mithen, gives us subtlety of humor.

Some kinds of cognitive disorder appear to be rooted in cognitive compartmentalization and a lack of fluidity. Autistic persons are often technically brilliant, able to recall enormous amounts of detailed data, but they normally have very routinized, channeled ways of thinking that do not allow for cross-fertilization of ideas from different domains of expertise.

Because humanity relies on invention to survive – again, we do not rely on physical adaptations, such as wooly fur, but on inventions, such as sealskin boots in the Arctic – humanity must be innovative and creative. This creativity demands and is driven by cognitive fluidity. The most creative people are those most able to connect thoughts from different fields; as Arthur Koestler wrote in *The Act of Creation*, creativity comes from "the sudden, interlocking of two previously unrelated skills or matrices of thought."[5] Cognitive fluidity, then, was selected for because it conferred the advantage of innovative thought, which in

---

[5]   Koestler (1964): 121.

turn aided in the refinement of the tools humans use to adapt to their environments.

Synthesis: One Mind, Two Models?

In this review we have covered two quite different explanations for the evolution of modern thought. For Donald, the modern mind evolved as an accumulation of novel modes of representation; ultimately, it was about the evolution of ways of voluntarily recalling detailed information from memory stores. For Mithen, the modern mind evolved as a consequence of a communication between previously-isolated modules of expertise; ultimately, it was about the evolution of innovation. Can these two very different explanations for the modern mind be reconciled? Mithen clearly embraces Donald's evolutionary approach to the modern mind, but feels that Donald did not make enough use of the archaeological record, and Donald has called Mithen's approach worthwhile, though he suggests that Mithen underestimates the significance of representation. It may well be that each model contains some truth, and in Figure 3.2 we synthesize the two models in one diagram. On the left, Mithen's cognitive fluidity increases over time, allowing fluid communication among different kinds of intelligence. Concurrently, in Donald's model (on the right), increasing specificity of memory recall leads to complex representations that are integrated into what cognitive scientist Liane Gabora has called 'worldviews'[6] that we might think of as distinctive cultures. To the right, we see that language, whether gestural and/or spoken, but certainly structured by a complex grammar, is used to communicate in both Donald and Mithen's models. Mithen's cognitive fluidity actively promotes new associations of previously-unassociated ideas; this is creativity, which is increasingly considered a measure of intelligence. Furthermore, Donald's increasingly specific and capable memory-management cognitive tools lead to a vast increase in the content of a given culture, archiving more knowledge at each generation, increasing the capacity for innovation and intelligence. Combined, these cognitive features allow a more specific fit of human action to selective environments, that 'fit' played out in proactive invention of specific cultural, technical and behavioral innovations that allow human survival in novel habitats. Increasingly through time, also, we see active niche construction – discussed further below – in which humans actively make ecological niches for themselves rather than simply adapting to what exists. Ultimately, it is the cognitive variation promoted by the features of language and intelligence that allows humanity – despite our relatively frail body – to colonize the globe. The lowermost panel in this diagram displays the traditional language groups worldwide, indicating something of the regional adaptation of human

---

[6]   Gabora (2006).

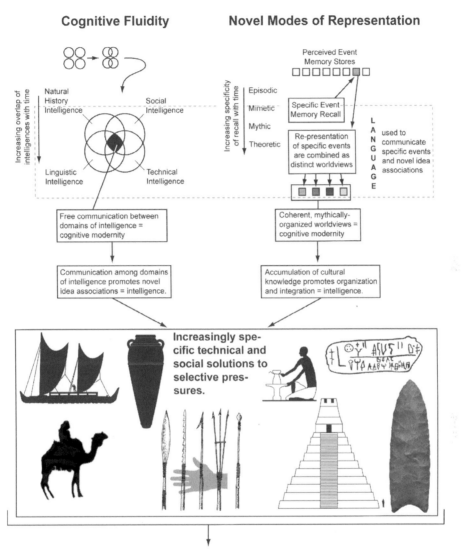

Figure 3.2. Synthesis of Mithen and Donald Models of Cognitive Evolution in the Genus *Homo*. Each model contributes to a better understanding of the coarse term 'intelligence', and each promotes novel idea association and increasing fit of human adaptations to their environments over time. Image by Cameron M. Smith.

cultures to environments worldwide, because language is strongly structured by the environment that it is used to describe and negotiate. Although this model remains provisional, it is a better description, we feel, than simply saying that 'humanity used culture and intelligence to colonize the globe'. This model allows a better understanding of the origin of regional cultures and human diversity than many such general statements.

Generally speaking, it is the mind of the genus *Homo* that allowed it to proliferate and spread as a species. All evidence today suggests that, while some variety of premodern *Homo* had emerged from Africa over 2.0 million years ago, the modern mind appeared first in African populations of *Homo* around 100,000 years ago. After that time, this cognitively-modern human species – also anatomically modern by this time – emerged from Africa and began to adapt to a multitude of environments across the globe. Figure 3.3 schematically displays this rapid, widespread adaptation, based on dozens of lines of evidence including archaeological sites and a recent proliferation of ancient DNA studies. In the figure, (1) shows the emergence of modern humans from Africa c.100,000BP, moving east across Sub-Himalayan Asia, and (2) shows their colonization of Europe beginning over 40,000 years ago. In colonizing Europe, the Neanderthals – a variety of *Homo sapiens* that had emerged from Africa earlier than modern humans, and had adapted to the conditions on the continent – were replaced by modern humans. In the same way, all other members of *Homo* that had adapted to different regions after the *first* migration of *Homo* out of Africa (over 2 million years ago) were replaced by the post-100,000 years ago emergence of modern *Homo*. In (3) we see the colonization of Australasia close to 50,000 years ago, and in (4) we see the crossing of the Bering Land Bridge and entry of humans into the Americas over 18,000 years ago. This is followed by (5), migration down the Pacific coast of the Americas by 15,000 years ago and (6) a later migration through the 'Ice Free Corridor' between the melting Cordilleran and Laurentide ice sheets (C and L, respectively). In (7) we see origin of the colonization of the Pacific Islands after 3000 years ago, and in (8) we see the rapid colonization of the high Arctic of North American and Canada, all the way to Greenland, mostly occurring after 1500 years ago. In these migrations and colonizations, humanity adapted to every conceivable environment, including windswept sea ice, open ocean, arid deserts, frigid plains, expansive wetlands, high mountains and both tropical and temperate forests.

It is critically important to remember that while other species also adapt to many different environments, their adaptations are largely biological, such that their anatomies diverge across geography, leading to subspecies as well as speciation, the formation of new species entirely. But in our genus, *Homo*, despite a few small anatomical adaptations to local geographical conditions, what diversifies radically is not anatomy and physiology but culture and behavior. This is why, as we mentioned earlier in this book, culture is so critical to understanding humanity; it is by culture that we survive, not anatomy, and in fact *despite* our relatively frail anatomy. In Figure 3.4 we see common body sizes and proportions in North American Arctic (on the left, A) and Nilotic North

Migrations in Prehistory

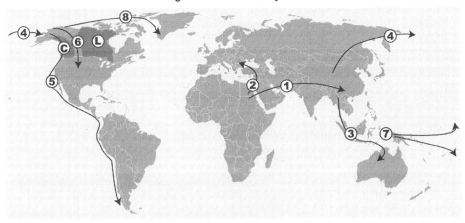

Traditional Language Groups

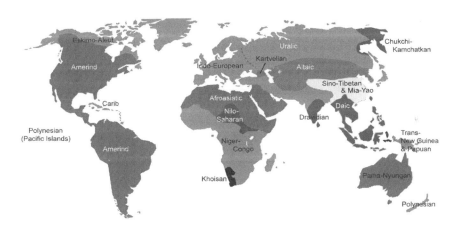

Figure 3.3. Human Colonization of the Earth Following 100,000 Years Ago. Image by Cameron M. Smith.

African (on the right, B) populations, showing phenotypic (bodily) variation in hair cover, skin color and body stature (see comments in Table 3.4) as adaptations to these environments. In Figure 3.4(C) we see material – artifactual – adaptations in terms of traditional hot-weather clothing of the Arabian peninsula, which includes light-colored, light-reflecting, loose flowing clothing that covers the skin but allows breezes to circulate freely. In Figure 3.4(D) we see similarly-critical artifactual adaptation in the form of clothing designed for the Arctic, with polar bear fur ruff around the face (which prevents wind from freezing the face), snowblindness-preventing goggles made of carved bone or wood, hand-preserving mittens, and sealskin boots with durable hide on the

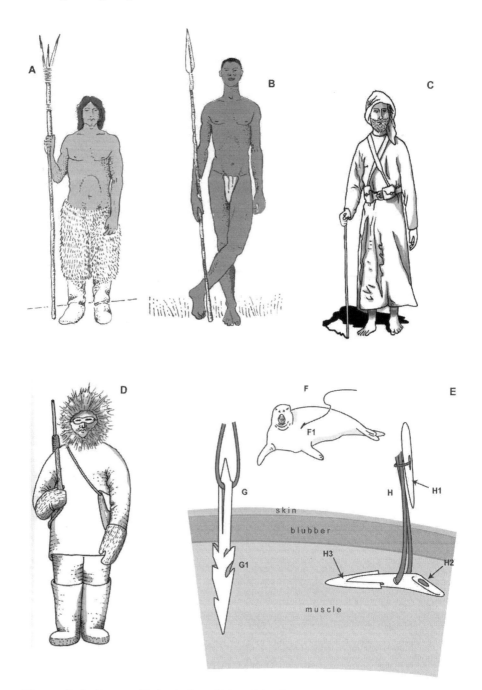

Figure 3.4. Human Biological and Material Adaptations to Hot and Cold Environments. Illustration by Cameron M. Smith, with special adaptation of parts E-H2 from Fagan (2000).

outside and insulating fur on the inside. In Figure 3.4(E) we see simple but highly effective Arctic hunting gear, designed for apprehending a sea mammal by securing it to a tether via an armature (F1). The tether normally leads to a thrusting weapon that is retracted immediately after harpooning. Figure 3.4(G) shows a simple harpoon armature imbedded in the animal; the barbs (G1) prevent the armature from working free, especially when the tether is stressed. Figure 3.4(H) shows the improvements of the *toggling harpoon* armature, which allowed rapid colonization of the high Arctic after 1500 years ago. In this design, a foreshaft (H1) is inserted into a socket (H2) in the armature. When the armature penetrates the muscle, the foreshaft is retracted, but the entire armature 'toggles' laterally, burying itself in the animal and making it less likely for the armature to pull free. An additional new element is the broad, wide, thin blade (H3) (often made of shell or slate) that is much sharper than a bone armature. With the armature toggled in the whale, walrus or seal body, it is much less likely to be torn out when hauling in on the retrieval cord.

In its rapid and wide spread across the globe, *Homo sapiens sapiens* devised adaptations such as those discussed above, as well as several main kinds of *subsistence*, each attended by a particular kind of social organization. These subsistence and organization modes tell us a lot about the variation of culture worldwide, and, as we will see, they tell us something very important for understanding human evolution. This is the lesson, which we will see in the next section, that there has been little inevitable about human evolution; certainly we did not move, predictably, from foraging to farming, as many believe.

Major Adaptations in Human Subsistence and Social Organization

In common with most animal life, we humans have some basic material requirements. Modern humans must meet the following basic needs on a daily basis, and our hominin ancestors must have approached some of the same figures:

- *Food* must be obtained; humans can metabolize fat, carbohydrates, and proteins and from some combination of these we must derive from between 1000 calories per day (a minimum for moderate health) to 5000 calories or more per day (to fuel an active person in a cold environment).
- About 2.5 liters (roughly half a gallon) of clean *water* a day for moderate health; much more might be required, depending on physical activity and atmospheric humidity.
- *Temperature regulation*; clothing, behavior and architecture must maintain a perceived temperature ranging from about 50F (10C) to 70F (21C) (below and above these temperatures, many humans begin to feel cold or warm, respectively).
- An assortment of *vitamins* and *minerals* must be obtained.

Table 3.1. Social Organization and Subsistence Among Humanity Past and Present.

	Band	Tribe	Chiefdom	State/Civilization
Subsistence	Foraging	Foraging/ pastoralism	Horticulture	Agriculture
Mobility	High	Medium/Cyclic	Low	Lowest
Food storage	Little: days to months	Little: weeks to months, or 'meat on the hoof' (among pastoralists)	Medium, seasons to a handful of years (some stored food crops)	High, with reliance on staple foods stored in large quantities for years at a time
Property ethos	Low but present.	Medium: among pastoralists, herded animals are property of individuals	Strong: elite classes own special objects unavailable to most of the population.	Very strong; highly ranked members own objects for bidden to those of lower ranks
Social ranking ethos	Little stratification; generally equal access to resources for all members	Medium: among pastoralists, families with more animals have higher rank	Strong: hereditary elite class exists, but has more power to coerce than command	Very strong: high rank can be achieved or ascribed, and access to resources depends on social rank
Population	10–150	Less than 200	Low hundreds to 1500	Tens of thousands to millions or billions
Example	Baka of Central Africa, Paiute of North American Great Basin, Inuit of Arctic Canada	Maasai of East Africa (cattle herders), Saami of Arctic Scandinavia (reindeer herders), Cheyenne of North American plains	Maori of New Zealand, Vikings of Medieval Scandinavia	Egypt, Greece, Shang (China), Maya (Mexico and Guatemala) United States

To fulfill these requirements as humanity spread into (and built) diverse ecological niches worldwide, every human culture devised a *mode of subsistence* that fulfilled these requirements (subsistence studies normally focus on calories, but we are slightly expanding the term here). Naturally, the way resources were distributed across landscapes presented options as well as constraints to behavior that in sum strongly conditioned subsistence strategies. Generally speaking, four main subsistence modes were 'settled into' by humans radiating out of Africa in the past 50,000 years. These are summarized in Table 3.1, and discussed below.

Foraging (also known as *hunting and gathering*) subsistence is characterized by high residential mobility, with human groups collecting their resources on a daily basis rather than relying on large, long-term stored resources (as among agriculturalists, mentioned below). The genus *Homo* has been practicing some variety of foraging for at least 2.5 million years, and tens of thousands of foragers still live today in the high Arctic, Southern Africa, Australia, and parts of South America. Although there are exceptions, important features of the general foraging subsistence mode include:

- High residential mobility, meaning that housing is normally impermanent.
- Limited food storage, meaning that environments or foods are unsuited to being stored in large quantities and/or for long periods.
- Lack of emphasis on possession, meaning that although some items may be individually owned, most are communally owned, and money (symbolic units of value) is absent.

Foragers are (and were in the past) most commonly organized as *bands*. These are relatively small groups, often composed of several nuclear families, who live and travel together. They know other bands, however, and exchange information and even genes with those others through marrying-in and marrying-out. Most foraging bands are essentially egalitarian, meaning that most resources are equally accessible to all members of the band.

Another common subsistence mode is *pastoralism*, the herding of animals for products such as milk, meat and wool. Pastoralists move their animals from one grazing patch to another, according to a complex seasonal cycle, so they are quite residentially mobile. There are tens of thousands of humans living with a pastoralist subsistence mode today, including the Samburu of Northern Kenya, who herd cattle, and the Saami of Arctic Scandinavia and Russia, who herd reindeer. Pastoralists do eat some meat, but rely more on their animals' *secondary products*, such as milk, butter, cheese, and hides, and in some cases the animals are also used to carry artifacts. Some key characteristics of the pastoralist subsistence mode are:

- Relatively high residential mobility (though less than among foragers) related to grazing-patch migration, meaning that housing is normally impermanent.
- Moderate degree of food storage, including 'meat on the hoof' in the form of living animals.
- Moderate emphasis on possession (compared to foragers), in which livestock are highly valued because they are labor-intensive to care for.

Pastoralist cultures, sometimes referred to as *tribes*, are normally somewhat larger in population than foraging bands. Tribes can have influential chiefs (absent among most foragers), but chiefs have more influence than actual power, and they can be kicked out by majority.

Like pastoralists, people practicing *horticulture* have domesticated other life forms, including both animals and plants; the plants are the focus, and crops are grown but their products are not stored for long periods due to preservation issues. Pacific islanders, traditionally, were horticulturalists, as are the Fore (for-AY) people of highland New Guinea, who focus on yams and pigs. Horticulturalists are somewhat 'tethered' to particular landscapes by their investment in it (ploughing) and the demands of farming, and they are significantly less mobile than foragers or pastoralists. Among the characteristics of horticulturalist societies are:

- High residential sedentism, staying in one settlement for generations at a time before moving (perhaps cyclically) to other farmland.
- Significant degree of food storage.
- Strong emphasis on possession of personal property, including farmland, food-processing facilities and food-storage facilities.

Many horticultural cultures were, in traditional times, organized as *chiefdoms*, such as those of the Maori of New Zealand. They were led by hereditary elites, people of a royal bloodline. Elites could not be as easily ejected tribal chiefs, and had considerably more influence. However, they could not extend political will very far, and while chiefs often built polities of up to several thousand 'citizens', such structures were unstable and normally disintegrated.

Humanity's fourth main subsistence mode has been *agriculture*, the intensive farming of domesticated plants and animals involving complex water-control facilities (including canals and dykes), intensive food-processing (including winnowing, separation of grain from chaff and grinding to powder before further processing), and reliance on stored foods that last for years after harvest and could be transported long distances as well. Important characteristics of agricultural cultures include:

- Total residential sedentism, staying in one settlement for generations at a time failing large-scale migrations.
- Reliance on stored foods to survive.
- Strong emphasis on possession of personal property, including farmland, food-processing facilities and food-storage facilities.

The first truly agricultural communities appear in the archaeological record in the Near East, around 10,000 years ago, but they appeared independently, worldwide shortly thereafter. Since this is the time of the end of the last ice age and the beginning of the current, warm period – known as the *Holocene* – it is presumed that climatic changes associated with the end of the ice age were somehow a driver for the invention of agriculture, but it is not clear how. Not all, but some agricultural cultures developed into *civilizations* or *states* characterized by a number of widely-recognized characteristics including tens of thousands to millions of members and populations with more (and more numerous) complex social and economic interactions with others of the population than found in chiefdoms, tribes or bands; these are summarized in

Table 3.2.[7] In chapter 4, *A Choice of Catastrophes*, we will discuss the collapse of civilizations in terms of the disintegration of some of the major structural elements of a given civilization; these structural elements are described by summary terms in the second column of Table 3.2.

Table 3.2. Essential Characteristics of Early Civilizations.

Characteristic	Summary term
Population densities greater than can be supported by immediate environment	*Urbanism*
Some members of culture completely disengage from food production and are fully employed as artisans	*Non-Food Production Specialists*
Reliance on agricultural stores for subsistence; taxation of domesticated products from citizenry	*Taxation and Tribute Agricultural Subsistence Base*
Use of permanent, highly-visible, costly structures as symbols of the supernatural domain and state power	*Monumental Architecture*
Centralized decision-making with delegation of administration to an hierarchy of permanent officers, involving political, military, and religious elites all led – most often – by a divine monarch	*Hierarchical, Centralized Authority, Social Ranking, Political Organization* and *State Religion*
Use of high-fidelity non-biological memory devices to record business, administrative events and religious concepts in great detail	*Durable Record-Keeping*
Increasing detail noted in workings of celestial bodies leading to investigation of natural cause and effect, and quantification	*Astronomy* and *Mathematics*
Elaboration of material culture used to signal social rank	Concept embedded in *Non-Food Production Specialists* (artisans)
Formalized trade routes with full-time traders and standardized trade-related units	*Long-Distance Trade Standardized Measures*
Relationships less organized by kin than by social position (class, status, rank)	*Social Ranking and Political Organization* (see also above)
Increased intensity, distance and duration of conflict	*Standing Armies & Territorial Sovergnity*

---

[7] These civilization characteristics are adapted from, and discussed further in, Childe (1950), whose seminal article in this topic remains of wide use in archaeology.

Reviewing these modes of subsistence and their attending social and economic organizations, it is critical to remember that anthropology has found no internal drive for humanity to single-mindedly move 'through' these stages, from one to the other. For example, agriculture simply will not work in the permafrost of the Arctic (or many other areas of the world), so in the Arctic people never developed agriculture or the highly populous, urban-dominated civilizations found where agriculture is practiced. Today, entirely cognitively modern human beings live in a foraging subsistence mode in many places worldwide, though their numbers are relatively low compared to the high populations of civilizations supported by a completely different mode of subsistence. The idea that all humanity would rise from one kind of existence to another – to use the archaic terms, from *savagery* (foraging) to *barbarism* (horticulture and pastoralism) to *civilization* (agriculture) – has been shown by a century of anthropology to be false. While in some areas precisely this trajectory *was* in the past followed (foraging did give way to horticulture, and then agriculture and civilization), the point is that it is not an inevitable, internally-driven 'ascent'. It is simply how subsistence modes – and their attendant cultural and political correlates – played out in various environments worldwide in the last 10,000 years.

With regard to human space colonization, we should keep in mind also that off-Earth cultures do not have to fit any of these molds in particular. While they will likely be supported by agriculture, and share many characteristics with those of modern, global civilization, elements that we might find very familiar – such as suburbs, or even cities – might not be present in all future off-Earth cultures. Ideas for how to shape those off-Earth populations, in terms of cultural and political structure, might well be productively informed by learning from how humans have arranged themselves in the past, in many different ways.

Niche Construction and Major Adaptations in the Evolution of *Homo*

For a long time, we in anthropology have established that humans adapt to the world in ways different from other life forms because of our intelligence. With a better understanding of what intelligence is, as we have reviewed above, we can specify that it is because of certain properties of human consciousness that humans adapt very quickly, precisely, and proactively. This is because culturally-accumulated knowledge stored in the mind – and outside the mind in external memory devices ranging from cave paintings to libraries – can be called upon to represent past events instantly and in extreme detail, allowing rapid, proactive tailoring of human actions and technologies to a given selective agent or environment.

We can also now invoke a relatively new body of theory to describe just how human adaptation is operationalized; this is *niche construction theory*. Niche construction theory addresses the world of proactive shaping of the selective environment by an organism. While it was proposed to explain non-human as well as human behavior, here we focus on how this body of theory pertains to human behavior.

Essentially, niche construction theory recognizes that many life forms are not simply passively-adapted precipitates of a given selective environment, but some behave in ways that alter the selective environment in ways that improve the fitness (a concept we reviewed in Chapter 1) of the organism and/or its offspring:[8]

> "Organisms do not just build environmental components, but regulate them to damp out variability in environmental conditions. Beavers, earthworms, ants, and countless other animals build complex artifacts [e.g. beaver dams], regulate temperatures and humidities inside them, control nutrient cycling and [chemical] ratios around them, and in the process construct and defend [suitable] environments for their offspring."

While niche construction theory has been somewhat controversial in that some consider it 'old wine in a new bottle' (only new terminology for what is already known[9]) we agree with archaeologist Bruce D. Smith that:

> "... human attempts at reshaping their natural landscapes [read 'selective environments' in the terms we use in this book] have been classed under a number of terms [including] 'indigenous management' ... 'domesticated landscapes' ... [and] 'indigenous resource management' ... all of these roughly synonymous terms fall comfortably now under the far more general heading of niche construction."[10]

Clearly, modern humans alter their selective environments in ways that improve their own fitness, as well as those of their offspring, and we agree that niche construction is a useful term and a useful body of theory (and a more considered terminology), for understanding human evolution. In fact, we are confident to state that the adaptation to off-Earth environments of whatever kind will be an unambiguous case of niche construction. Table 3.3 summarizes the significant properties of and differences between niche construction and natural selection. (Table 3.3 is adapted from Table 4.1, p.176 of Odling-Smee, Laland and Feldman (2003).)

In short, today anthropology can more precisely say how it is that our physically frail body was carried, by a complex, intelligent mind, to the multitude of environments we inhabit today. It is important to remember that the bulk of the many ecosystems we occupy today were not adapted to with modern, 'high' technologies; even Neanderthals occupied the bitterly-cold

---

[8]   Laland and Brown (2006): 95.
[9]   See Brodie (2005): 249.
[10] Smith (2007): 191.

Table 3.3. Comparison of Niche Construction and Natural Selection.

Niche Construction (NC)	Natural Selection (NS)
Sources of NC are active, 'fuel-consuming' agents; living organisms	Sources of NS are either abiotic environmental components (e.g. temperature or rainfall regime) or other niche-constructing organisms
NC must be advantageous in the short term, aiding the organism's survival and reproduction	NS need only obey laws of physics and chemistry
NC must usually be restricted to fitness-enhancing behaviors or processes in the short term. NC is unlikely to be random or haphazard	NS has no goal and while progress and directionality can arise, they are not intended
NC is directed by active choices made by individual organisms	NS is passive; 'it' is not a coherent process at all, but simply differential survivorship of variable members of a population of life forms
NC is largely proactive, in that it establishes conditions for the future	NS is, in essence, reactive, in that life forms are born with characteristics that worked in *their parents'* time; NS cannot not 'look' into the future

Russian steppe over 40,000 years ago. Rather, most of our basic adaptations were accomplished in prehistory with relatively simple technologies, complex social arrangements that buffered individual human beings from many selective pressures, and a few biological adaptations. And we humans clearly modify our environments, as seen in Figure 3.5, the arrangement of irrigation canals on the Nahrwan River in Southwestern Iran. The river is the thickest gray line, and major and minor canals (black lines) branch away from it, leading to agricultural fields. Many lead directly to settlements, shown here as filled circles. These settlements and canals reflect not one moment in time, but a period of over a century after 900 AD.

Table 3.4 summarizes some of the many adaptations used by humanity in three particularly challenging environments, distinguishing between biological and cultural and technical solutions to limiting factors of certain environments. This is an illustrative prelude to the close look we will take, in Chapter 6, at the colonization and adaptation to the Pacific environments, and in Chapter 7, at how to shape the human colonization of off-Earth environments in ways that are intelligently informed by an understanding of human evolution to date and in principle. (Table 3.4 is based on Moran (1979) Table 1.1 with other information derived from throughout Bajema (1971), Jurmain, Kilgore and Trevathan (2011), Palmer et al. (1999), and Roberts (1973).)

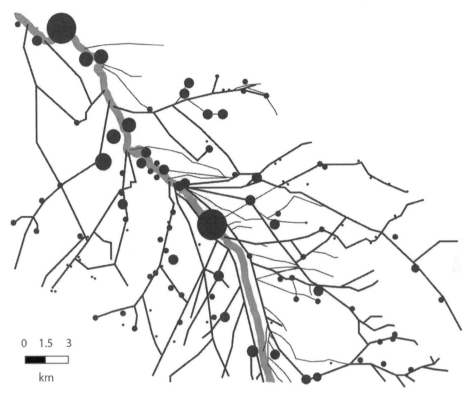

Figure 3.5. Niche Construction as Seen in Irrigation Facilities, Southwestern Iran. Image by Cameron M. Smith, adapted from Fig. 8-3 of Jacobsen and Adams (1966).

Table 3.4. Limiting Factors and Adaptations of Modern Human Populations to Arctic, High Altitude and Arid Environments.

Biome	Limiting Factors	Biological, Cultural and Technological Adaptations
Arctic/ Cold	• Extremely low temperatures for long periods • Extreme light/dark seasonal cycles • Low biological productivity	Biological: • increased Basal Metabolic Rate • increased shivering, vasoconstriction and cold thermoregulation activity and efficiency • compact, heat-retaining body stature Cultural & Technological: • bilateral kinship = demographic flexibility • clothing insulates but can prevent sweating • semisubterranean housing including igloo made of local, free, inexhaustible resource (snow)

Table 3.4, cont.

Biome	Limiting Factors	Biological, Cultural and Technological Adaptations
		• high fat diet yielding many calories and vitamins • low tolerance of self-aggrandizement • low tolerance of adolescent bravado • high value of educating young • social fission • mobile, field-maintainable, reliable tools • population control methods including voluntary suicide and infanticide • high value on apprenticeship • low tolerance for complaint; 'laugh don't cry'
High Altitude	• Low oxygen pressure • Nighttime cold stress • Low biological productivity • High neonatal mortality	Biological: • dense capillary beds shorten distance of oxygen transport • larger placenta providing fetus with more blood-borne oxygen • greater lung ventilation [capacity] Cultural & Technological: • promotion of large families to offset high infertility • use of coca leaves to promote vasoconstriction and caffiene-like altertness • woolen clothing retains heat when wet • trade connections with lowland populations
Arid/Hot	• Low and uncertain rainfall • High evaporation rate • Low biological productivity	Biological: • tall, lean, heat-dumping body • lowered body core temperature • increased sweating efficiency • lower urination rate • increased vasodilation efficiency Cultural & Technological: • flexible kinship system = demographic flexibility • intercourse taboos maintain sustainable population • loose, flowing clothing blocks sunlight • wide sandals block ground-reflected sunlight • nakedness accepted during physical labor

Adaptations of the Genus *Homo*: Overview and Lessons for Human Space Colonization

As we saw in Chapter 2, it is possible to step back from the details of evolution and observe larger trends. In the case of the evolution of our genus, *Homo*, we argue that three major adaptations have so far occurred in the past 2.5 million years, and that the next – the extraterrestrial adaptation, characterized by the permanent colonization of space – would be a natural continuation of expansion and other trends seen in the previous human adaptations; human evolution has been characterized by both biological and cultural adaptations to new environments, largely via niche construction. Table 3.5 summarizes the chief distinctions of biological and cultural adaptation.

Broadly speaking, there have been three fundamental adaptations in human prehistory. The *Terrestrial Adaptation* occurred roughly five million years ago, when our ancestors became habitual bipeds in a profound biological adaptation to African savannahs (by two million years ago, our ancestors reached China). The *Technological Adaptation* began around 2.5 million years ago, when we began to rely more on technological adaptations (such as stone tools and fire) than biological adaptations. The *Cognitive Adaptation* occurred around 100,000 years ago, characterized by the advent of abstraction, which changed the way we processed information about ourselves and our environment. By 50,000 years ago we possessed modern consciousness (discussed in the next chapter), and our adaptations included highly complex interpersonal communication (language) and the management of cultural information, including symbolism. The Cognitive Adaptation eventually led to farming, urbanization, writing, and mathematics. These adaptations are what allowed our ancestors to move beyond Africa and flourish in regions of the globe that, earlier in our evolution, would have been impossibly hostile. We should remember that despite its capacity for adaptation, our genus might well have had a number of 'close shaves' in prehistory; Figure 3.6 illustrates a proposed population crash of *Homo* around 70,000 years ago, after a massive eruption of the Toba volcano in South-East Asia, which some consider to have caused global temperature to drop far below even those at the time of the glacial maxima (coldest periods) of the Ice Age; another, unrelated population crash is suggested around 20–25,000 years ago.[11]

We propose that the most profound forthcoming adaptation will be the *Extraterrestrial Adaptation*, the movement of humans into space colonies where viable populations will exist biologically independent of Earth populations.

The more we know about how human adaptation works, the better the chances of succeeding with the Extraterrestrial Adaptation. Table 3.5 summarizes the key characteristics of biological and cultural modes of adaptation.

---

[11]   See Ambrose (1998).

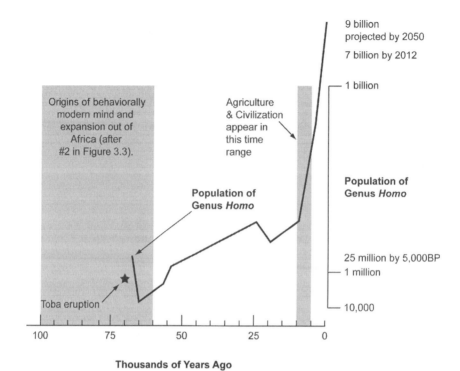

Figure 3.6. Population Crash of Genus *Homo* Proposed After the Toba Eruption Around 70,000 Years Ago. Illustration by Cameron M. Smith after Figure 5 of Ambrose (1998).

Table 3.5. Adaptive Characteristics of Biological and Cultural Evolution.

	Transmission of Information	Timing of Adaptation	Specificity of Adaptation	Potential for Rapid and Specific Adaptation to Selective Pressures and Environments
Biological	Sexual, via DNA, to next generation only = slow and narrow[1]	Based on reproductive schedule = slow	New traits unlikely to specifically 'fit' selective pressure.	Low
Cultural	Social, via language, to any peers or any members of any future generation = potentially very rapid and wide	Can spread instantly = fast	New traits can be tailored specifically to selective pressures.	High

[1] this excludes horizontal gene transfer, discussed elsewhere in this book.

In Chapter 6 we provide a compelling example of human adaptation and niche construction from human prehistory, illustrating that while technology has figured largely in human adaptation for millennia, that technology has not been the focus or main concern of our adaptive endeavors. The focus has been people looking for a new place to live, using an ingenious array of adaptations to very different environments.

Three key lessons applicable to human space colonization can be derived from this examination of the evolution of life and of the genus *Homo*.

First, as we found with the evolution of life at large, the survival of our genus was not inevitable; the Neanderthals, for example, became extinct after 200,000 years of flourishing in Europe and the Near East. Again, then, we must be very careful of the outdated phrase and sentiment that 'human destiny is in the stars'. An examination of the contingencies of evolution has shown that destiny, in fact, has no scientific meaning except in the broadest sense of a life form's genetic heritage, and to rely on this concept ignores the proaction that has characterized human adaptation to many environments, not passively, but proactively, using language, information-rich, instantly-retrievable stores of past events to guide our invention of adaptive tools, both behavioral and cultural. We must also remember that human cognition, consciousness and intelligence are unique on the Earth and worth preserving, and our better understanding of what intelligence itself is will be a tremendous asset. We must continue to be proactive in our adaptations and we must favor and value cognitive flexibility because it aligns the operations of the mind with the fundamentally dynamic, rather than fixed, nature of the universe. Space colonization will be more easily accomplished if its technologies and techniques are designed to address such change, and that can only be done if the minds designing the technologies are encultured to embrace change rather than fixity.

Second, evolution and adaptation are often expansive, exploratory phenomena, leading to the inhabitation of every imaginable – and some unimaginable – environment and econiche by some form of life or another. Though not goal-directed or consciously driven in any particular direction except when directed by *Homo*, this expansion and exploration is a fundamental property of life, which spreads, even out from the original location of sessile (non-locomoting) life forms such as plants (Figure 3.7 shows the dispersal of spores from fungal plants, in which both tall and short forms release spores not into the laminar air current near the ground, but the more turbulent air flow, where spores are more likely to be carried farther away.) The continued exploration of space, then, is in fact – and as so many feel – a deeply natural, evolutionary process, and a life-affirming one. This is important to recognize because much of the activitiy of humans in space so far has involved inaccessible and often secret technologies that have alienated many people and even turned people away from the concept of human space colonization. But these opaque and exclusive elements of humans-in-space so far are not a necessary result of the use of technology, any more than the use of stone tools,

Figure 3.7. Spore Dispersal from Ground Fungi. Illustration by Cameron M. Smith.

over two million years ago, was an alienating event. Humanity has expanded and explored, and found new homes across the Earth. The Earth is now, largely, filled with humans, and all our eggs are in this single planetary basket. To survive – as we will see in the following chapters – we must spread out, and doing so would be in keeping with all of Earth's evolutionary history. To not do so, in fact, would be to resist evolutionary history and Nature, and that, it seems, never works out, at least in the long run. We must continue to construct new niches for human habitation. It will be natural to do so off of Earth, not unnatural.

Third (although this was not discussed in detail), the human colonization of the globe in prehistory was at first accomplished by explorers, but the game soon changes from exploration to colonization, which means, in short, having plenty of offspring to build genetically viable communities. Thus we have precedent to consider the short-term, goal-oriented philosophy of human space exploration so far – often understandable as humanity developed the technologies to explore space – as essentially different from and incompatible with the larger goals of human space colonization. The themes, we will explore later in this book, that motivate these different projects (exploration and colonization) differed significantly in the past as much as they do today. For this reason, we argue that the movement of space access away from exclusively government-funded efforts is a significant, positive movement. Tying space exploration to taxpayer dollars and political whims will not, in the long run, result in successful space colonization. A significant component – perhaps the bulk of it – will have to be done privately, and with a philosophy of seeking a *beginning* in space, rather than a finite conquest. We must have long-term goals, such as the establishment of new, Earth-independent populations of our species – rather than focus on short-term exploratory goals.

How to implement this continued evolution? How do we transform 'outlandish' ideas, such as space colonization, into reality? We will return to this important point later in this book. First, however, it is necessary to examine in detail the arguments that have been forwarded for and against the human colonization of space. We will turn to these in the next two chapters, before returning to the topic of how to use the lessons of adaptation at large, and of our genus and species in particular, to guide and shape human space colonization as an adaptive endeavor.

Part II

Arguments For and Against Human Space Colonization

4 A Choice of Catastrophes: Common Arguments for Space Colonization

"The furnaces of Pittsburgh are cold; the assembly lines of Detroit are still. In Los Angeles, a few gaunt survivors of a plague desperately till freeway center strips, backyards and outlying fields, hoping to raise a subsistence crop. London's offices are dark, its docks deserted. In the farm lands of the Ukraine, abandoned tractors litter the fields: there is no fuel for them. The waters of the Rhine, Nile and Yellow rivers reek with pollutants."[1]

"Let me not seem to have lived in vain"

Tycho Brahe, dying, to Johannes Kepler in 1601.

Most human cultures have doomsday myths, ranging from Biblical stories of Armageddon to the Scandinavian tales of Ragnarok – the "Twilight of the Gods". Hindu cosmology holds that the Universe ends in fire and regenerates from water. Ancient Persian Zoroastrians believed along lines similar to Judeo-Christians – that an ultimate savior would be born and rid the world of evil – although for the Zoroastrians this involved the Earth and its inhabitants perishing in a flow of molten metal before renewal, just as Aztec cosmology proposed cyclic annihilation and renewal. Doomsday prophecies are also common in secular culture, such as the urban myth of 'Planet X' (also known as 'Nibiru'), anticipated by many (though without any evidence) to impact the Earth with calamitous force in 2012.[2] These myths constitute something of a cultural universal; they cross temporal and cultural boundaries with the recognition of limits – even if only periodical – to human existence at large.

---

[1] From "The Worst is Yet to Be?", *Time* magazine, 24 January, 1972.

[2] Planet X (or Nibiru) is popularly rumored to be positioned to collide with Earth in late December 2012; many note that this date coincides roughly with the end of the current cycle of the ancient Maya calendar, though at least some traditional Maya shamans have stated that such a calamity is *not* predicted. The 2009 Columbia Pictures film "2012" is an example of a recent "end of the world" film, as are the recent films "Armageddon" (Touchstone) and "Deep Impact" (Paramount). Some, of course, try to bring about the end by their own means – such as the 1995 Sarin gas attacks in the Tokyo subway system by members of the Aum Shinri Kyo 'Doomsday' cult – though they have always lacked the technical capacity to do more than local harm.

These limits are not entirely mythological. Even if humanity were to end war, overpopulation, disease and pollution, ensure global justice and build a network of defenses against such cosmic dangers as solar eruptions and wandering comets and asteroids, the Sun cannot be prevented burning out, at which time its plasma shell will expand and incinerate the Earth and all human works. The Sun's expansion is not expected to occur for another five billion years, and may be thought of in a somewhat mythical way. But there are certainly serious and immediate threats to the human species that, we argue, make a compelling case for beginning the migration from Earth sooner rather than later.

We are not the first to point these out, of course; in his 1979 book *A Choice of Catastrophes*[3] Isaac Asimov discussed a variety of plausible natural and culturally-caused events that could cause the extinction of humanity, or at least collapse global civilization. While humanity has taken action on some of these threats – for example, an international effort now scans the sky for 'civilization-killer' comets and asteroids[4] – many of Asimov's proposed calamities could still occur today. Unfortunately, some are more likely today than in the past, such as the use of nuclear, chemical or biological weapons by individuals or small organizations, and the already-apparent effects of global over-consumption of natural resources, which defense organizations worldwide already recognize as likely leading to resource wars in the relatively short term.

Asimov made many of these points nearly 40 years ago, but more recent surveys of the possibility of relatively near-term human catastrophe have been published, and they are not encouraging. A context for these projections has been forwarded by philosopher Robert Heilbroner, who has argued in the book *Visions of the Future* that from the time of early humans to the 17th century AD, most of humanity saw its future as essentially changeless in its material and economic conditions, a position that paints with quite a broad brush. Perhaps more perceptively, he also argues that from the 18th century AD to the mid-1900s, Western civilization (at least) saw its future as essentially bright and positive, to be achieved through the application of science, whereas since the mid-1900s (significantly, after two World Wars and the invention of nuclear weapons) there has been a more varied conception involving negatives resulting from "impersonal, disruptive, hazardous and foreboding" factors,[5] though including some positive hope.

Technology figures large in these conceptions, and it is clear that science and the technologies that derive from it can yield great opportunities as well as terrible risks. These were important issues to Asimov, and are more important today. A recent review by Oxford University philosopher and futurist Nick Bostrom points out that three recent discussions of the near human future by prominent thinkers have highlighted significant threats to human existence

---

[3]   Asimov (1979).

[4]   NASA Near Earth Object Program: http://neo.jpl.nasa.gov/.

[5]   Heilbroner is paraphrased in Bostrom (2007): 7.

within the next 1–5 centuries; John Leslie gives humanity a 30% chance of becoming extinct in the next five centuries, Astronomer Royal Martin Rees has weighed in with a figure of a 50% chance of extinction within the next 90 years, and Bostrom himself giving humanity a greater than 25% chance of extinction in the next century. Of course, these are speculations, but they are informed speculations and they reflect technological and other realities that could not have informed earlier, mythical doomsday concepts we discussed above.[6]

Natural threats to humanity include impacts on Earth from extraterrestrial objects such as asteroids and comets. Human-caused threats to humanity, or at least civilization (defined and discussed in Chapter 2), include ecological overexploitation and conflicts using nuclear, biological and/or chemical weapons. The magnitude of threats to humanity range widely (e.g. from extinction to substantial reduction of the species population); we focus on the levels of (a) the extinction of *Homo sapiens sapiens* or (b) the collapse of modern civilization.

Extinction

Extinct species are those whose members have all died out; they may be known to humanity in the fossil and/or DNA record of ancient life forms, but are no longer living at present. Humanity has only been scientifically aware of the 4.5-billion-year age of the Earth for about 100 years, and for much of humanity's more recent history we have considered Earth to be a relatively safe and benign home, at least between cyclic catastrophes. But palaeoenvironmental and fossil records show that calamities and extinctions have been common through time. In a comprehensive survey of the paleontological record paleontologist David Raup has documented that over 99% of all species that have ever lived on Earth have become extinct, and that most species (e.g. *sapiens*) have a duration of about four million years, while most genera (e.g. *Homo*) have a duration of about 20 million years.[7] While these are fascinating figures, we must recall that, as we will see through this book, such figures apply to life forms that do not know they are evolving in the first place, and can therefore do nothing proactively about significant threats to their selective environments – their habitats. Humanity, as we saw in Chapters 2 and 3, however, is unique in its ability to both perceive such changes and, if time allows, adapt to them. We return to this important point at the end of this chapter.

Extinction normally takes place over multiple generations; millions of generations for faster-reproducing species, thousands for slower-reproducing species. It often results from changes in selective environments that are too rapid for a given species to adapt biologically. For example, when a comet (or asteroid)

---

[6]   Bostrom (2007): 10.
[7]   Raup (1991).

struck the Earth around 65 million years ago , selective environments changed due to the cloud of debris that was spewed into the atmosphere; the cloud blocked sunlight, which caused changes in temperature, vegetation regimes and so on. This was a change of selective environment so rapid that dinosaurs were unable to adapt with the biological evolution of novel traits suitable to their new selective pressures. Species can also become extinct if they are out-competed by other life forms that are more proficient at life in a given selective environment, as when North American mammals migrated south and replaced many South American marsupials, starting around 3 million years ago.

The history of life on Earth includes several well-documented *mass-extinction* events in which large percentages of Earth life – or some segment of Earth life – became extinct. These events are so distinctive in the fossil record that the disappearance of an established life form and the appearance of new one in the paleontological record are often used to define the beginnings and ends of the geological periods. Such events could occur again and it is clear that most would either cause human extinction at least the collapse of modern civilization.

Some mass extinctions occurred over millions of years due to gradual changes in the environment, and some – as in the well-known comet or asteroid impact that ended the reign of the dinosaurs – occurred, from the perspective of life form adaptation, instantly. In each case, full recovery of the Earth's biodiversity took tens of millions of years. We will examine some such extinction events after considering another possible scenario: not extinction, but civilization collapse.

Civilization Collapse

Civilization, as we saw in Chapter 3, is an arrangement of human social, economic, religious and other relationships, all supported by an agricultural subsistence base. We also saw that it is a pervasive myth that civilization is an end-point in human evolution; it is one way of being human, and other ways – such as living in small foraging cultures – have existed for far longer than any ancient civilization. Having said this, we do not advocate humanity's abandoning civilization for another mode of life. Despite the deep problems with modern civilization, we feel it its advantages are worth preserving (at the same time, we should not expect all humans to inhabit civilization *per se*, and it is of course the right of the few human cultures that continue to forage, herd animals, or practice horticulture as their way of life, as among modern people of the high Arctic, parts of Africa and Eurasia, and New Guinea, respectively, for example).

While it is widely felt that civilization is the end-point in human evolution, that illusion is dispelled by a quick examination of the collapse of ancient civilizations through the past few thousand years. Precisely what constitutes the collapse of a civilization is debated in archaeology and anthropology, but archaeologist Joseph Tainter has made a compelling case, based on a wide survey of the literature, that civilization collapses are characterized by:[8]

- A reduced degree of social differentiation.
- Less economic and occupational specialization of individuals, groups and territories.
- Less centralized regulation of diverse economic and political groups.
- Less behavioral control and regimentation.
- Less investment in the epiphenomena of complexity: monumental architecture, arts, sciences etc.
- Less flow of information between groups and individuals and between center and periphery.
- Less sharing, trading and redistribution of resources.
- Less coordination and organization of individuals and groups.
- Smaller territories per political unit.

Figure 4.1 presents a chronology of ancient civilizations, with indications of their many collapses.

Generally speaking, the results of civilization collapse are population dispersal, decentralization of decision-making and the administration of justice, and economic independence of settlements, rather than economic interdependence. How does this play out, materially? In the case of Maya civilization, agricultural decline was attended by the decentralization of agriculture, the material backbone of all civilizations ancient or modern. In our own civilization, agricultural collapse could also result in a demographic shift of populations away from urban centers (where most people live today) and out to 'hinterlands' where they could farm locally to support themselves. While it is arguable that some of these (and other) post-collapse conditions might in fact be improvements over modern civilization, it should be remembered that in post-collapse times, local overlords have nearly always arisen in the power vacuum, building their own quasi-military forces that have been used to control the larger portion of the population, who become rural peasants, as in the case of feudal Europe. For the late Belgian historian and expert on the European Medieval period, François-Louis Ganshof (1895–1980) feudalism is characterized by the following traits:[9]

- Weakened state with inability to protect its territory or interests.
- State unable to protect citizenry.
- Endemic collusion of remaining administration and powerful social agents.
- Endemic use of personal (e.g. family) relationships in remaining political and economic systems.
- Prevalent elitism, restricting opportunities of most of the population.
- Use of mercenaries for security and protection.

---

[8] Tainter (1988): 4.
[9] Ganshof (1964).

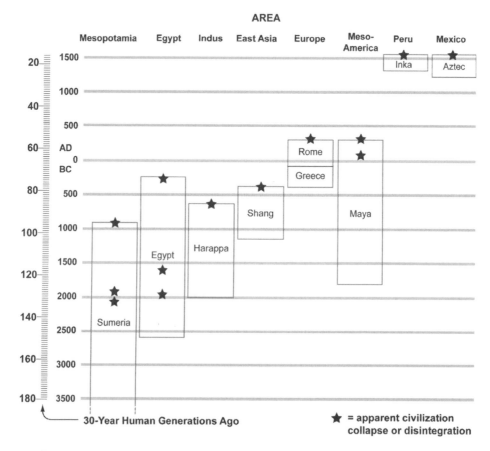

Figure 4.1. Chronology of Ancient Civilizations, Indicating Collapses. Image by Cameron M. Smith.

Civilizations have collapsed (perhaps a better word is 'disintegrated') to such conditions many times, as we will see shortly. What has caused such collapses? The precise reasons have been many. Tainter indicates several, including warfare, ecological and/or vital resource depletion, political disunity (stemming from mismanagement and/or mistrust among the populace), natural catastrophe, insufficient response (adaptation) to altered ecological circumstances, climate change (today's concern), and others, all of which result in sharply declining reduction of the effectiveness of administrative complexity,[10] while Jared Diamond, in a less scholarly (but more popular) treatment, has emphasized ecological collapse as responsible for many ancient civilization collapses.[11]

---

[10] Tainter (1990).
[11] Diamond (2005).

Tainter's hypothesis, that civilizations collapse when the costs of maintaining the system outweigh the benefits, is based on a more subtle reading of the archaeological evidence, and is perhaps more convincing. In his model, collapse happens when the state can no longer support its own structures, often economically, as in the case of Rome, where military over-extension resulted in an inability to pay troops and other government officials that maintained the system. Whatever the case, it is clear that there are many paths to collapse; it could result from material (e.g. ecological) or data-management (e.g. administrative complexity) issues, or, of course, both of these, and other features. And the result could be any kind of decentralization, on a spectrum including the feudalism mentioned above. While some civilizations have recovered from ancient collapse, others have not.

Modern civilization at large can be criticized for ignoring the past and focusing on the present; this results in the illusion, mentioned above, that modern civilization is a natural end-point to human evolution, and that it is somehow invulnerable to collapse. Of course, members of ancient civilizations must also have considered collapse unthinkable, as in the case of any Roman or Egyptian citizen at the height of empire. But with the perspective of archaeology we see the reality. Table 4.1 presents our analysis of the collapse of nine ancient civilizations, and places our own in this context.

In Figure 4.2 we see the archaeological traces of the collapse of Mycenaean (A) and Maya (B) civilizations, respectively. In the case of Mycenae, we see a rapid reduction in the number of occupied Mycenean settlements over three centuries; since no settlements grew appreciably through this time, this reveals population decline and/or movement away from traditional settlements to previously-unoccupied 'hinterlands'. And in the Maya case, a collapse in a New World civilization, we see that the number of occupied settlements (per 20-year 'kahun' or Maya political counting period) drops dramatically after the mid-700s AD, also reflecting a dispersal of the population.

By the reckoning of Table 4.1, there have been 377 generations of civilization on Earth in the last 6000 years (457 if one counts Western civilization), with one generation being an average of 30 years. The average duration of a civilization has been 42 generations, or 1260 years. Of nine ancient civilizations, four ended (by our classification) in conquest, three by ecological decline that undermined the agricultural backbone, one by a slow withering and replacement by a rival power (Greece) and one by the costs of military over-extension and invasion from outside (Rome) (since many civilizations ran concurrently, the fact that 377 × 30 = 11,310 years – and the earliest civilization begins around 6000 years ago – is not an error but only reflects that the generational experience of civilization is not sequential.) On a larger scale, if we consider human existence at large to begin with the evolution of the genus *Homo*, at about two million years ago, and we use a 30-year generation time (this would have been shorter in some times, and longer in others), another way to say this is that there have been about 67,000 generations of human experience, and by rough calculation we see that only 0.6% of that experience has been in a mode – civilization – that most of us

Table 4.1 Duration and Collapses of Ancient Civilizations.

Civilization	Duration	Reasons for Collapse	Comment
Sumeria	6000BP – 2900BP = 3100 years = 103 generations	Ecological decline (soil salinization) leading to depopulation	Area later repopulated
Egypt	5100BP – 2400BP = 2700 years = 90 generations	Conquest by Persians	–
Harappa	4500BP – 3300BP = 500 years = 16 generations	Ecological change (altered rainfall regime) leading to depopulation	–
Shang-Zhou	3600BP – 2700BP = 900 years = 30 generations	Conquest by Xirong	–
Maya	3800BP – 1000 BP = 2800 years = 93 generations	Ecological decline leading to conflict	–
Aztec	800BP – 490BP = 310 years = 16 generations	Conquest by Spain	End is 1521 AD
Inka	600BP – 479BP = 121 years = 4 generations	Conquest by Spain	End is 1532 AD
Greece	2800 BP – 2156 BP = 644 years = 21 generations	Slow replacement by Roman civilization	End is annexation of Greece by Rome in 146 BC
Western Rome	2400 BP – 1500 BP = 900 years = 30 generations	Overextension, invasion, political corruption	End is considered at 476 AD with decentralization; Eastern Rome (Byzantium) continues until 1453 AD
Modern Western Civilization	2400BP – present = 2400 years = 80 generations	–	Origin is considered Greece of c.400 BC

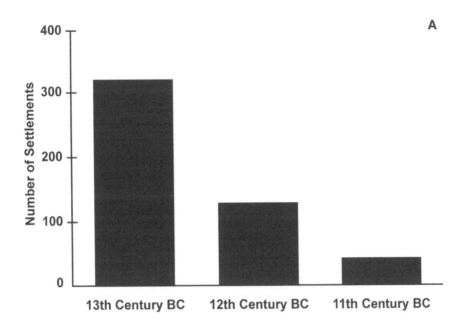

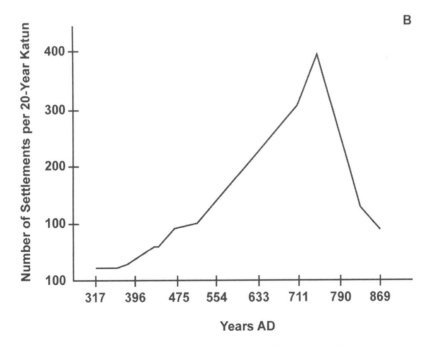

Figure 4.2. Collapses of Mycenean and Maya Civilizations. Images by Cameron M. Smith based on Tainter (1999).

would recognize. That the greater bulk of human experience has been lived as small, mobile, foraging cultures, followed by only about 10,000 years (about 333 generations) of life as pre-civilization agriculturalists. It is important to remember that while the average duration of a civilization has been about 1200 years, or about 40 generations, this figure is not useful as a predictor because civilizations have collapsed for many reasons, often unpredictably.

What are the lessons, then, that we can take from this examination of the collapse or disintegration of ancient civilizations? The first is that civilization is a very new experience for humanity, and its continuation has less to do with a proven track record or inevitability than with our own incredulity that it could actually end. The second lesson is that all ancient civilization have experienced collapses, and that while many have reassembled after collapse, the periods in-between have not been utopians free of centralization, but essentially Medieval in their conditions, without what we consider the benefits of civilization, including libraries, centers of learning, and freedom from subsistence concerns for many individuals, allowing their engagement in efforts such as special craftsmanship or space colonization. A third lesson is that while collapses of ancient civilizations might have been isolated – as in the case of Rome collapsing with no effect on the contemporary New World civilizations (such as the Maya) – the interconnection of modern civilization suggests that such independence is a past luxury. Economist Robin Hanson has pointed out that today's global civilization is "... so highly specialized and interconnected that a disruption in one ... could create a cascading series of disruptions in all of them."[12] This is clear enough, but it could also be argued that interconnection actually adds resiliency (capacity to rebound from disturbance) to complex systems, including global civilization. In the same way, Tainter has argued that the kinds of civilization collapses experienced in the past are less likely in the modern world because:

> "Every nation is linked to, and influenced by, the major powers, and most are strongly linked with one power bloc or the other. Combine this with instant global travel, and, as [social philosopher] Paul Valery noted, ' ... nothing can ever happen again without the whole world taking a hand.' ... Collapse today is neither an option nor an immediate threat. Any nation vulnerable to collapse will have to pursue one of three options: (1) absorption by a neighbor or some larger state, (2) economic support by a dominant power or by an international agency; or (3) payment by the support population of whatever costs are needed to continue complexity, however detri-mental the marginal return. A nation today can no longer unilaterally collapse, for if any national government disintegrates its population and territory will be absorbed by some other."[13]

---

[12]   James Martin 21st Century School, *Policy Foresight and Global Catastrophic Risks* (2008): 4.
[13]   Tainter (1988): 213.

While it is good to remain optimistic, and to not be unduly motivated by fear, it is possible to imagine realistic scenarios in which the essential elements of interconnection of modern global civilization are significantly severed, if not resulting in a complete collapse, at least resulting in so much chaos and expense in rebuilding that other, longer-term priorities – such as working to ensure that humanity protects itself from Earth catastrophe by making an adaptation to off-Earth environments – would be put aside. Such disasters have visited the Earth before, though sometimes so long ago that humanity has only learned about them with new, scientific methods. To ignore what we have discovered about the illusion of safety on our home planet is to gamble with the survival of our species. In the next section, we review some ancient disasters that, had they occurred more recently, could well have caused human extinction, or at the least, the dissolution of civilization, with no guarantee that it would ever recover.

Natural Threats to Humanity and Civilization

Of the natural threats to the integrated systems of modern civilization, perhaps the most serious – in their capacity for large-scale damage, ranging from human extinction to civilization collapse – is the threat of the Earth being impacted by a comet, asteroid or other space object. As the Richter Scale estimates the damage from various magnitudes of earthquake, the Torino Scale has been established to classify the threat posed by various space objects that could impact the Earth:[14]

0: The likelihood of a collision is zero, or is so low as to be effectively zero. Also applies to small objects such as meteors and bodies that burn up in the atmosphere as well as infrequent meteorite falls that rarely cause damage.

1: A routine discovery in which a pass near the Earth is predicted that poses no unusual level of danger. Current calculations show the chance of collision is extremely unlikely with no cause for public attention or public concern. New telescopic observations very likely will lead to re-assignment to Level 0. Merits mention by astronomers.

2: A discovery, which may become routine with expanded searches, of an object making a somewhat close but not highly unusual pass near the Earth. While meriting attention by astronomers, there is no cause for public attention or public concern as an actual collision is very unlikely. New telescopic observations very likely will lead to re-assignment to Level 0.

3: A close encounter, meriting attention by astronomers. Current

---

[14]   Adapted from Morrison et al. (2004).

calculations give a 1% or greater chance of collision capable of localized destruction. Most likely, new telescopic observations will lead to re-assignment to Level 0. Attention by public and by public officials is merited if the encounter is less than a decade away.

4: A close encounter, meriting attention by astronomers. Current calculations give a 1% or greater chance of collision capable of regional devastation. Most likely, new telescopic observations will lead to re-assignment to Level 0. Attention by public and by public officials is merited if the encounter is less than a decade away.

5: A close encounter posing a serious, but still uncertain threat of regional devastation. Critical attention by astronomers is needed to determine conclusively whether or not a collision will occur. If the encounter is less than a decade away, governmental contingency planning may be warranted.

6: A close encounter by a large object posing a serious but still uncertain threat of a global catastrophe. Critical attention by astronomers is needed to determine conclusively whether or not a collision will occur. If the encounter is less than three decades away, governmental contingency planning may be warranted.

7: A very close encounter by a large object, which if occurring this century, poses an unprecedented but still uncertain threat of a global catastrophe. For such a threat in this century, international contingency planning is warranted, especially to determine urgently and conclusively whether or not a collision will occur.

8: A collision is certain, capable of causing localized destruction for an impact over land or possibly a tsunami if close offshore. Such events occur on average between once per 50 years and once per several 1000 years.

9: A collision is certain, capable of causing unprecedented regional devastation for a land impact or the threat of a major tsunami for an ocean impact. Such events occur on average between once per 10,000 years and once per 100,000 years.

10: A collision is certain, capable of causing global climatic catastrophe that may threaten the future of civilization as we know it, whether impacting land or ocean. Such events occur on average once per 100,000 years, or less often.

Currently NASA's little-known Spaceguard Survey monitors space for Earth threats, as summarized on NASA's *Asteroid and Comet Impact Hazard* website:

"The Earth orbits the Sun in a sort of cosmic shooting gallery, subject to impacts from comets and asteroids. It is only fairly recently that we have come to appreciate that these impacts by asteroids and comets (often called Near Earth Objects, or NEOs) pose a significant hazard to

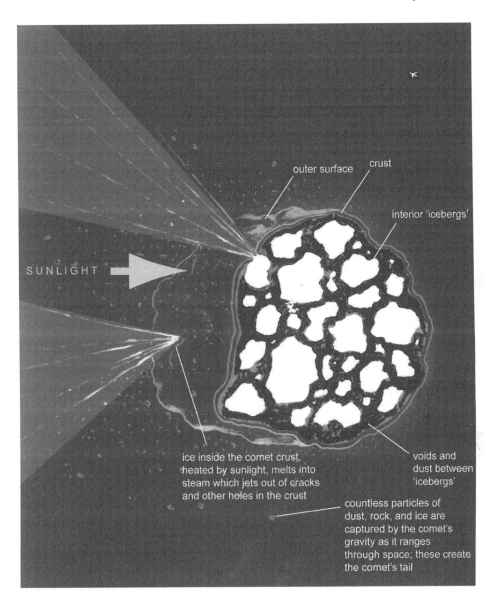

Figure 4.3. Schematic View of a Medium-Small Comet. Note that a Boeing 747 has been superimposed on the upper right as a measure of scale. Image by Cameron M. Smith.

life and property. Most of the hazard is from asteroid impacts; comets make only a minor contribution. Although the annual probability of the Earth being struck by a large asteroid or comet is extremely small, the consequences of such a collision are so catastrophic that it is prudent to assess the nature of the threat and prepare to deal with it ... Today we are addressing this impact hazard by carrying out a comprehensive telescopic search for potentially hazardous near-Earth asteroids (NEAs) ... [the Spaceguard Survey] has already resulted in the discovery of [many] NEAs larger than 1 km diameter."[15]

The extinction potential of such space objects – and other natural threats to humanity and civilization – become stark when we see what science has to say about them. In the following section, we review the main Earth extinction events known to science. That not all resulted from space object impact should remind us that catastrophe could come from more than one direction. Figure 4.3 is a schematic depiction of a medium-small-sized comet; note that a Boeing 747 has been superimposed on the upper right to really give us an understanding of the scale.

The End-Ordovician Mass Extinction

In rocks approximately 445 million years old, geologists find marked changes in carbon and oxygen isotopes, along with the dramatic reduction of fossils of brachiopods, corals, trilobites and other genera. Between 450 and 440 million years ago up to 60% of the Earth's marine invertebrates died off, marking the end of the Ordovician period and the beginning of the Silurian.

While extreme volcanism and even an atmosphere-destroying flood of gamma rays from a supernova within a nearby arm of our Milky Way galaxy have been considered as possible causes for the End-Ordovician mass extinction, most available evidence suggests that the drifting of the Gondwana supercontinent over the Earth's south pole resulted in global glaciation. Figure 4.4 illustrates the position of the continents and how the Earth's surface appeared at this time. This early ice age appears to be associated with a dramatic lowering of sea levels, drying out vast habitats of shallow-water life. Since it is known that today shallow-water species typically form the bottom of the food chain, supporting all other life, it is argued that the end-Ordovician mass extinction was caused by the collapse of the shallow-water food chain base, leading to all higher members of the ecosystem – such as the plant-eaters and the flesh-eaters – to also collapse.

Since the Ordovician/Silurian collapse was probably very slow, perhaps comparable to modern climate change, humanity could probably survive a comparable event, even though there could be large-scale starvation. On the

---

[15] Bostrom (2007): 10.

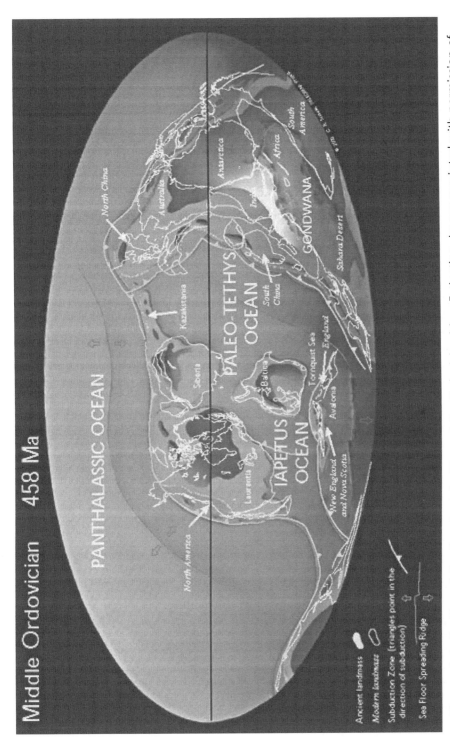

Figure 4.4. Position of Continents at the Time of the Ordovician Mass Extinction. Image reprinted with permission of C.R. Scotese, PALEOMAP Project, Arlington, Texas.

other hand, more sudden change to the oceanic food web could be devastating to the human population. If the base of the food web (e.g. oceanic krill) collapses, the upper elements (e.g. all that feed on krill and all that feed on those that feed on krill) collapse; this means the failure of fisheries, and today well over half of humanity obtains the bulk of its protein from marine sources. Could oceanic food webs collapse catastrophically? Recent research has shown a very worrying 40% reduction in oceanic phytoplankton – the base of the oceanic food web – in the last century; not only that, but other research has shown that every fishery discovered in the last 500 years has been systematically fished to the brink of collapse.[5]

The Ordovician/Silurian collapse appears to have occurred due to sea-level drop, but sea level *rise* is also a significant danger to, at least, the stability and wellbeing of global civilization. Documented rising global temperature is widely accepted to be causing sea level rise by melting of Antarctic, Greenlandic and non-polar (e.g. European alpine) glacial ice. In 2007 the Intergovernmental Panel on Climate Change (IPCC) estimated that by the year 2100 there would be a 48–79 cm (approximately 19.2–31.6 inch) rise in mean global seal level.[16] A catastrophic breakaway and melting of the West Antarctic Ice Sheet alone could raise sea level by 5–6 meters (approximately 17–20 feet). Similarly, the Greenland ice sheet contains enough ice to raise sea level about 7 meters (approximately 23 feet) which would displace of millions – perhaps billions – of people worldwide as they would be compelled to move to higher ground from the urbanized coastlines. There would also be vast expenditure in the construction of stop-gap solutions such as seawalls; and the Greenland ice sheet is already melting.[17]

In either slow-rising or slow-lowering sea levels with effects comparable to the Ordovician/Silurian event, changing coastlines and their effects on marine ecosystems can most probably be coped with technologically. Human populations would be stressed and would have to spend vast sums to relocate or develop new food sources, but the gradual changes are not apparently factors that could cause human extinction. Sudden changes, on the other hand, could be catastrophic and undermine the bases of modern civilization. Were the Greenland and the West Antarctic Ice Sheet to melt rapidly, even over the course of a decade, there would be enormous pressure placed on humanity in terms of floods of refugees, the cost of evacuations and the need to distribute foods, medicine, fresh water and other supplies, not to mention the inevitable

---

[16]   The International Panel on Climate Change (IPCC) cautions however that rate of sea level rise may change over time, and that all the mechanisms responsible for sea level rise remain poorly understood.

[17]   There is a small chance that the West Antarctic ice sheet could collapse within a few centuries, but the response of the ice sheet to future climate change is uncertain and a subject of debate. The Greenland ice sheet is already contributing to sea level rise (from melting), but it does not contain the same instabilities as the West Antarctic and most projections suggest a gradual melting (IPCC, 2007).

conflicts. Such an event could distract humanity for decades or longer. Still, neither scenario suggests either the extinction of the human species or the explicit collapse of industrial civilization, although tremendous suffering and waste could result.

"The Great Dying": The End-Permian Extinction

The largest mass extinction event we have evidence for occurred approximately 251 million years ago, at the end of the Permian period and the beginning of the Triassic: the end-Permian event. This event also marks the end of the Paleozoic era and the beginning of the Mesozoic era, the nearly 200 million year reign of the dinosaurs. Figure 4.5 illustrates the position of the Earth's continents at this time, when the continents were joined as the supercontinent known as *Pangaea*, surrounded by the super-ocean *Panthalassa*. As the continents drifted into one another there were massive volcanism events, and at the same time, the paleontological record shows that many new varieties of life were evolving. As Professor Richard Lund of the Carnegie Museum of Natural History told us, "It was a very busy time in Earth's history."

While the end-Permian event is the largest extinction event known, it happened so long ago that the geological record is unclear regarding its causes. Still, most paleontologists agree that 57% of all families of life – and 83% of all genera – died out during this time, equating to an extinction of over 50% of all marine species and an estimated 70% of land species. Brachiopods, mostly chitinous inarticulate burrowing benthic animals, and corals, suffered severe losses during this time. Additionally, many marine mollusk species, especially those that relied on shells rich in calcium carbonate, became extinct.[18] As one paleontologist has put it, the end-Permian event was the "most severe biotic crisis in the history of life on Earth".[19] More prosaically, we can call it a very close shave for Earth life.

What caused the end-Permian mass extinction? Mounting evidence indicates that the super-ocean Panthalassa was becoming anoxic (deficient in dissolved oxygen to the point where it could not sustain life) towards the end of the Permian. This is reflected in massive deposits of anoxia-related marine sediments near what is today Greenland,[20] as well as variations in uranium/thorium ratios in several late-Permian sediments wordlwide.[21] Panthalassa as a whole, then – not just in the region of Greenland – was severely deficient in oxygen. But what caused oceanic anoxia? Two main theories have been forwarded. Traces of super-craters in both China and Australia have been interpreted as the results of the

---

[18] Shin & Shi (2002).
[19] Kaiho et al. (2001).
[20] Wignall & Twitchett (2002).
[21] Levin (2010): 236.

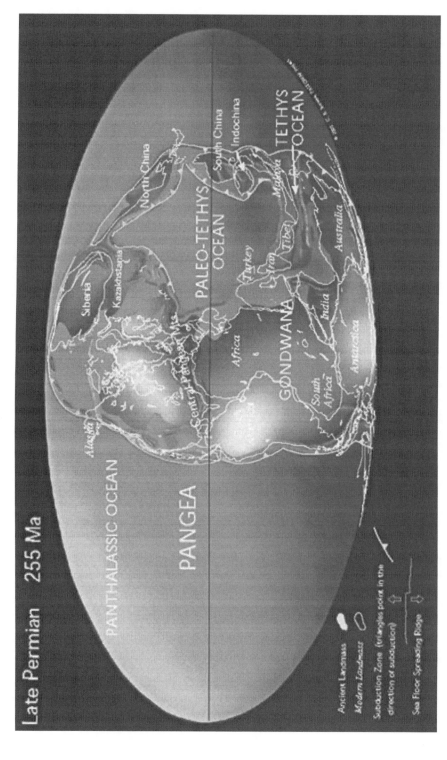

Figure 4.5. Position of Continents at the Time of the End-Permian Mass Extinction. Image reprinted with permission of C.R. Scotese, PALEOMAP Project, Arlington, Texas.

impact of fragments of a massive space body (perhaps a meteor or a comet) that vaporized and exploded upon contact with the Earth's surface, resulting in titanic plumes of vaporized crust, rich in carbon, ejected from the Earth's mantle and then settling in the seas to produce oceanic anoxia.[22]

Alternatively, and perhaps alarmingly, some of the most recent evidence points to another culprit. The dramatic increase in Pangean-related volcanism, according to some researchers at the University of Calgary, suggests that massive volcanic eruptions burnt significant volumes of coal and produced ash clouds that had broad and similar impacts on global oceanic oxygen levels.[23] More critically, such large volumes of carbon being forced into the atmosphere would have led to a massive worldwide greenhouse effect, whereby solar radiation enters the Earth's atmosphere, becomes trapped, and superheating occurs. Therefore, a major culprit for the greatest mass extinction ever known on Earth is something ecologists talk about with regard to our society everyday: global warming.

Oxygen isotope data indicates the temperature of equatorial ocean waters had risen at least $6°C$ by the end of the Permian, and fossil evidence indicates that many families of vertebrates perished as warmer conditions moved into their habitats. The volcanism that produced the Siberian Traps, one of the largest basalt flood plains in the world (covering over two million square kilometers) would have released tremendous amounts of CO_2 into the atmosphere, increasing the greenhouse effect and warming the Earth. In the Permian oceans as well, the Earth was choking on carbon. Data from sediments that were deep under the sea in the Permian period reveal that extremely high levels of CO_2 were persistent in ocean water. As the CO_2 circulated through the ocean, it probably poisoned many marine organisms.

More critically, warmer ocean temperatures would also have significantly reduced the amount of dissolved O_2 that the water could contain in solution, basically causing asphyxiation of a multitude of species, very similar to the effects of warmed water pumped into streams and rivers by factories. Pollutants aside, the water temperature alone makes the water less capable of supporting life. As an added effect of the Permian global warming trend, Geologist Harold Levin notes that large quantities of methane gas frozen in sea floor sediments may have been released, "driving global warming to even more disastrous levels."[24]

The global warming theory for the Permian extinction event is supported by the timeline for end-Permian mass extinctions that probably occurred over a period of perhaps as much as 1–2 million years. What is beyond question is that the world's selective environments changed so severely that most Earth life was

---

[22] Becker et al. (2004).
[23] "World's Biggest Extinction Event: Massive Volcanic Eruption, Burning Coal and Accelerated Greenhouse Gas Choked out Life" *Science Daily* Jan. 25, 2011.
[24] Levin (2010): 376.

unable to adapt, even over the scale of two million years. One would expect because of the long stretch of time involved that most species should have been able to react and adapt; to survive. But many did not. Whatever their source or time-scale, the changes that occurred in the end-Permian environment were so severe that they led to the extinction of most Earth life. At the end of the Permian our planet was nearly dying, its ecological make-up becoming unable to support fewer and fewer kinds of life.

While super-volcanism of the sort that probably occurred at the end of the Permian is very unlikely to occur today, a similar effect could be generated by our own doing. Through our industry we are unquestionably putting significant amounts of carbon into the atmosphere, causing a warming trend. We must pay very close attention to these trends, and take all possible steps to reduce the amount of carbon we put into the atmosphere. If we were to suffer an event that ultimately killed only half the number of species – as in the Permian – the ensuing competition for dwindling resources and foodstuffs would probably trigger global warfare that would disintegrate civilization long before humanity starved to death. Once, in the past, it seems, global warming – from one source or another – was responsible for almost ending all Earth life.

The End-Cretaceous Mass Extinction

Another major extinction event was also rapid, and appears to have been triggered by a single event, one that is common enough in our Solar System. This is the impact of one celestial body on another, in this case the impact of an asteroid on the Earth, with cataclysmic consequences. The end-Cretaceous extinction, securely dated to about 65 million years ago, is perhaps the best understood of the Earth's mass-extinction events.

Until the late 1970s extreme volcanism, climate change, and the onset of an ice age were all proposed to explain the end-Cretaceous event, but it was in 1980 that Nobel-prize winning father and son team Luis and Walter Alvarez discovered a very thin layer of iridium in Italian sediments dated precisely at the 65 million year mark referred to as the "K/T boundary".[25] While iridium is rare on Earth (0.4 parts per million) it is quite abundant in meteorites (470 parts per million) and abnormally abundant in terrestrial sediment from the K/T layer (6 parts per million). This simple fact gave scientists their first hard evidence that the end of the dinosaurs may have been related to an impact of an extraterrestrial object. The second shoe dropped when Canadian geologist Alan Hildebrand collected strong evidence that an ancient crater discovered by oil prospectors in 1978 near the town of Chicxulub (the 'x' is pronounced 'sh') on Mexico's Yucatan peninsula was contemporary with the end-Cretaceous event, as well as

---

[25]   Alvarez (1997).

abundant samples of "shock metamorphism" and closely-spaced planar fractures considered diagnostic of geological metamorphosis caused by a significant impact.[26]

Although there are a few holdouts, today the majority of the scientific community agrees that a bolide impact caused the Chicxulub crater 65 million years ago, and that that impact was primarily – if not solely – responsible for the end-Cretaceous mass extinction. Studies of the Chicxulub crater allow geologists to estimate the size of the asteroid that impacted the Earth: 10km (6 miles) in diameter and resulting in explosive force on the order of 250 billion kilotons of TNT; the force of approximately 1.6 *billion* atomic bombs (the Little Boy atomic bomb dropped on Hiroshima, Japan, on August 6, 1945, exploded with an energy of about 15 kilotons of TNT). As Robert Zubrin points out, "... the question is no longer what killed the dinosaurs. The real mystery is how anything else survived."[27]

As with the possible bolide impact responsible for the end-Permian mass extinction, the end-Cretaceous impact is suspected to have thrown such a tremendous amount of light-blocking debris into the atmosphere that many plants could not photosynthesize; as they died, so did the herbivores that dined on them, and the carnivores that dined on the herbivores, and so on up the food chain. In the evolutionary terms we have seen in this book, we see that these species' selective pressures were so radically and rapidly altered that there was no chance to adapt. Not all life was extinguished – far from it – but the 200-million-year history of the dinosaurs came to an abrupt end.

There is evidence that at the end of the Cretaceous there was a period of renewed volcanism that would have released tremendous amounts of dust, aerosols and sulfuric acid into the atmosphere, which would have blocked solar radiation and caused temperatures to decline. In turn, the sulfuric acid would have generated acid rain that could have changed the pH of the oceans enough to place lethal stress on plankton and other marine organisms, having a reverberating effect up the food chain. There is also evidence that there was a global lowering of the sea level at the end of the Mesozoic, perhaps brought on by seafloor spreading rates. As we recall from the Ordovician/Silurian extinction, this spells disaster for the marine life that depends on the shallows near the continental shelf, and by extension all the higher predators that depend on them. Some geologists suggest that with the spreading of the continents the tempering effect of the epicontinental seas the land masses would have been subject to more extreme climate conditions.

Perhaps the K/T extinction resulted from a combination of all three of these events. The cooling conditions and lowering of the sea levels (perhaps caused by cooling conditions) could have been produced as an effect of the decades-long

---

[26]   Levin (2010): 458.
[27]   Zubrin (1999): 131.

"nuclear winter" that would have followed an impact by a 10km-wide asteroid. Whether it was solely responsible for the end of the dinosaurs, the asteroid impact certainly put terrible stress on a great number of organisms. Moreover, the conditions on Earth after the impact of such a massive space object were certainly not helpful for any living thing. What happened on that last, lazy afternoon in the late Cretaceous went something like this (and it could happen again):

As the asteroid entered the atmosphere it would have created a devastating supersonic shock wave that would have superheated the air in the middle and lower atmosphere and converted it to plasma, incinerating anything in the way. Once the white-hot asteroid actually splashed down, it would have vaporized rock, blasted immense quantities of the Earth's crust into space as ejecta in addition to creating a boiling "supertsunami" wave that would have flooded vast areas of dry land. The ejecta thrown into space by the blast would then itself reenter the Earth's atmosphere at terrific velocities, raining super-hot ejecta which would then have set fire to forests all over the Earth, with the resulting gas, dust and ash in the atmosphere darkening the sky, creating poisonous acid rain and blotting out the sun for years at least. Once the sunlight did return, the immense quantities of CO_2 thrown into the atmosphere by the impact would have created a significant greenhouse effect and a subsequent global warming event. The remaining land plants, their numbers still low from the impact event, could not quickly eliminate this surfeit of carbon, and the post-Chicxulub Earth was probably very inhospitable for a long time after the impact. Still, many kinds of life did survive, and it is after the extinction of the dinosaurs that we see a proliferation of relatively new kinds of life, including early mammals.

It is difficult to imagine our relatively fragile civilization today, supported by an agricultural backbone, surviving such an event. While pockets of humanity would probably survive the impact event, it is hard to imagine how even they could survive such a 'cometary winter'. As we find ourselves presently confined to one planet, the human story could very easily come to an end.

In 2008, a study by researchers at the Cardiff Centre for Astrobiology in the United Kingdom revealed that as our Solar System passes through the densest part of the galactic plane there are times when we are more prone to asteroid strikes than others. So serious a threat is the impact of a large space object (generically referred to as a *bolide*) that in 1995 NASA established the Near Earth Object Program specifically to monitor comets and asteroids that have the potential to impact with the Earth. According to the space agency, "As of February 13, 2011, 7817 near-Earth objects (NEO) have been discovered. Some 822 of these NEOs are asteroids with a diameter of approximately 1 kilometer or larger. Also, 1199 of these NEOs have been classified as Potentially Hazardous Asteroids (PHAs)."[28] Here is a typical entry, taken from April 8th, 2010:

---

[28]   See http://neo.jpl.nasa.gov/risk/.

"A newly discovered asteroid, 2010 GA6, will safely fly by Earth this Thursday at 4:06 p.m. Pacific (23:06 U.T.C.). At time of closest approach 2010 GA6 will be about 359,000 kilometers (223,000 miles) away from Earth - about 9/10ths the distance to the Moon. The asteroid, approximately 22 meters (71 feet) wide, was discovered by the Catalina Sky Survey, Tucson, AZ"[29]

What we would do in the event of forewarning of an impact is another matter. We do not currently possess the level of technological sophistication to alter an asteroid's path or destroy one completely. A common misconception is that all we would need to do to prevent such an impact is launch a specially designed missile at the object and destroy it. Even if such missiles existed (and they do not) breaking up an asteroid would do us no good, as the smaller pieces could simply regroup due to gravitational force such that we would be struck by all of the fragments.

Our best hope lies in early detection, but more importantly in the study of these objects. As John Lewis, Professor of Planetary Sciences at the University of Arizona writes, "these objects will certainly collide with Earth. The only uncertainty is when these devastating impacts will occur".[30]

To summarize, the end-Ordovician event collapsed oceanic biota, which today would collapse fisheries. The impact on the food chain would be enormous, and would result in tremendous loss of human life. It is not clear however that humanity would become extinct in such an event, even if it were relatively abrupt. Sea-level rise, or decline, associated with such an event or resulting from climate change, would also cause tremendous suffering, but probably would not cause human extinction, though it might be so costly that it undermines and destabilizes modern civilization. And the end-Permian extinctions certainly appear to have been brought about by massive warming and climate change, probably ultimately by the impact of a space-object with the Earth. But even the collision with a smaller object that might not directly cause human extinction would certainly cause global famine and turmoil, and would almost certainly cause the destabilization and possibly the collapse of modern civilization.

The lesson here is stark: albeit on long time-scales relative to human lives, generations or political careers, space debris and dramatic climate change are probably significant threats to the human species and certainly to the integrated and in some ways rather fragile modern civilization. Despite our modern consideration of the Earth as a safe home, in the longer view it is not necessarily the safest basket to contain, as many have pointed out, all of humanity's eggs. Ecologists and paleontologists have shown that surviving extinctions events is much more likely if a species is numerous, widely distributed, and biologically diverse, points we will return to later in this book. Currently, humanity is

---

[29] See http://www.jpl.nasa.gov/news/news.cfm?release=2010-115
[30] Lewis (1997): 7

numerous – on the order of seven billion people – but, in terms of solar-system calamities we are very restricted in its habitat, to the planet Earth, and while we are genetically and behaviorally diverse, we could increase those qualities by fostering new human populations off of Earth.

Extinction of Humanity vs Extinction of Civilization

We have seen how mass extinction events have led to the demise of a great many life forms, and have illustrated how the same events could and in some cases certainly would have the same effect on the human species if they were to occur today. There is however an additional factor to consider regarding human extinction. As we pointed out in Chapter 1, humans are cultural animals; that is to say that for the last million years or so, we have relied on our tools and our society to adapt to the world around us rather than (primarily) our biology. Humanity survives today not so much because our bodies can endure the selective pressures of the myriad habitats we occupy worldwide, from the Arctic to the Congo, but because of the voluminous and intricate cultural information that we have accumulated that tells us how to build artifacts that provide us with shelter (allowing *thermoregulation* to keep us cool or warm), clean water, food and nutrients, and so on, in these diverse environments. So long as some humans do that, the species is alive; but we are arguing not just for the survival of the species, but the survival of modern civilization. What if there were an occurrence or series of events that did not cause human extinction, but instead collapsed our civilization?

As we have shown, ancient civilizations collapsed repeatedly. The ancient civilizations were largely independent, however, so if one were to fall – such as the Maya around 900 AD – others could persist, such as the contemporaneous Chinese (Shang). A significant difference between these ancient civilizations and our current civilization is that today, unlike ever before in the past, all industrial nations are to some degree interdependent, such that significant collapse or disintegration in one area could affect humans worldwide. What would that world look like? It might be similar to the European Medieval period, which followed the collapse of Rome and was characterized by feudal landholding, feuding one-upmanshnip of small competing states, a lack of centralized (secular) decision-making, information storage, and information processing, a lack of formalized education, intercultural languages and information flow, and an emphasis on the past – rather than innovation – as a guide to the future.[31]

Such a scenario sounds like a modern doomsayer's prophecy, but the collapse of modern civilization could occur more easily than we realize or like to think. This could result from natural or human-induced changes.

---

[31]   Grabois (1980).

Human Threats to Humanity and Civilization

Civilization Collapse: Where Are the Maya?

The Maya ruins of Tikal, Palenque, Copán, Chichen Itza and many others are magnets for the curious. Scholars, tourists and school children come from various parts of Mexico and indeed from all over the world to view the structures left behind by the ancient Maya. Even those with only a passing interest in Mexico's pre-Columbian past cannot help but wonder when gazing at the ancient but sprawling urban development; the temples and palaces, the magnificent pyramids and ball courts; where did the people who build all of these things go? What happened to the Mayan kings and their armies, the high priests, the astronomers, the scribes, architects and the artisans who carved blocks of limestone stone into magnificently intricate stelae?[32]

Put this way, any archaeologist, or Maya for that matter, would feel that the question need not even be asked; the Maya are still there. They still speak their own language and incorporate their own cosmology into their lives and even practice their traditional medicine. The Maya did not go anywhere at all.

The fields of the Yucatán are still worked by Maya people; towns and villages are still populated by Maya farmers, but after the fall of classical Maya civilization around a thousand years ago there were no more Maya architects to build monumental temples or palaces, and the sculpting of monumental hieroglyphic records of dynastic history ceased. Knowledge gained over a millennium of astronomical observations – sophisticated enough to predict eclipses – was forgotten. The architectural and engineering ability gained through centuries of practice and experimentation could not be taught to new apprentices without the formal system and infrastructure support (quarries, masons, builders) of a civilization. There were no apprentices to teach; everyone was engaged in food production. All of the scientific and technical grandeur of classical Maya civilization was lost, and many the epicenters of their glory were abandoned. Much like the European Dark Ages, the ability to read and write the Maya script persisted in some pockets of the population, but illiteracy became the norm. With the arrival of the Spanish conquistadores, the remaining Maya books were burned. This comprehensive collapse likely took place in matter of a few generations. Archaeological evidence indicates that there was a large-scale exodus of people from the urban centers in the central lowlands of the Yucatán peninsula around 900 AD. While some smaller settlements survived in the north of the Yucatán and managed to reconstitute themselves somewhat by 1250 AD,

---

[32] Maya *stelae* are tall sculpted stone columns, usually inscribed with glyph writing, depicting the divine lineage of the king and glorifying his deeds. They served as historical markers but had many other functions, not all of which are known. Many ancient civilizations constructed similar monuments.

they existed only as competing city states and never recovered the civilization-level integration of Classic Maya civilization.[33]

What caused the collapse of the classic Maya civilization? The immediate causes are debated, but it is clear that centralized government failed, and along with it the redistribution systems that provided food and shelter for the skilled classes, the non-food production specialists (introduced in Chapter 3), which included scribes, architects, astronomers and the multitude of other non-farmers who support the integrated efforts of all civilizations. The organized redistribution networks – the Classic Maya economy in essence – is what allowed the architects and the astronomers and the scribes to pursue their arts and sciences, excusing them from subsistence activities. In the same way that today non-food production specialists – like us and most of our readers – buy food from supermarkets and grocery stores, live in houses or apartments that someone else built, drive cars that someone else designed, and fill them with fuel that someone else refined, the non-food production specialists of the Maya (and those of other collapsed civilizations) for some reason could no longer be supported.

Archaeologists have long investigated the reason for the Maya collapse, generally dated to about 900 AD. Theories range from over-farming of the land, which would have led to less productive crop yields and created a chronic and worsening food shortage, to theories of climate change involving a severe 200-year drought and even cult activity. Professor Michael Coe of the Peabody Museum of Natural History at Yale University states "Almost the only fact surely known about the collapse was that it really happened."[34]

It appears that several factors played a role, and not insignificant among them was overpopulation coupled with food shortages induced as a result of destruction of the environment. Archaeologists have increasingly discovered evidence for massive deforestation in the Maya region: Coe notes "by the end of the eighth century the Classic Maya population … had probably increased beyond the carrying capacity of the land. In short, overpopulation and environmental degradation had advanced to a degree only matched by what is happening in many of the poorest tropical countries today. The Maya apocalypse, for such it was, surely had ecological roots."[35]

In the Maya and Angkor cases we see likely evidence of changes to the selective environments of the domesticated species that fuelled the civilization-level demands of human cultures. Certainly Maya people still exist today, but the civilization from which they are descended collapsed, fragmented, dispersed and disintegrated due to ecological changes that reduced food surpluses to the extent that non-food production specialists could not be supported and those people had to physically move away from urbanized centers to farmable plots of land

---

[33]   Coe (1993).
[34]   Coe (1993): 128.
[35]   Coe (1993): 127.

that could support smaller populations. This quite literally dis-integrated the civilization elements – introduced in Chapter 3 – of the Maya civilization.

The critical support of non-food production specialists seems distant in modern civilization, where we often purchase and consume food far from where it was produced, but every calorie consumed by non-food production specialists – from engineers to accountants – must come from a viable agricultural system. Such systems, as we have seen, are globally vulnerable to changes resulting from agents as disparate – and possibly uncontrollable – as climate change and space-debris impact with the Earth. Dramatic Earth impacts or super-volcanism might not be enough to cause human extinction, but they may well break the agricultural backbone of modern civilization. The same can be said of less-dramatic, but just as dangerous, climate changes that could radically alter the selective environments of the domesticated species humanity relies on today to survive; a handful of staple plants (rice, wheat, corn and beans) and a handful of animals (cows, pigs, chickens and fish) that themselves subsist on the plant crops.

Splendid Isolation: The Great Leap Backward of Ming Dynasty China

Perhaps the best known Chinese navigator and a great explorer in his own right was Cheng Ho, also spelled Zheng He, the Admiral of the Ming Dynasty fleet, revered to this day as China's "Christopher Columbus."[36] In the early 15th century, Ming Dynasty China was arguably the richest, most influential, most knowledgeable and most complexly-organized society in the world. The ships of the Ming navy were remarkable for their time and they may in fact have been the largest wooden sea-going vessels ever built. While there is some dispute as to their actual size, accounts by figures such as Marco Polo and the Arab geographer Ibn Battuta claim that some of the ships in the fleet may have been almost 450 feet long and over 100 feet wide in the beam. Some of these ships were reported to have had between five and nine masts and carried compliments of between 500 and 1000 passengers.[37]

Between about 1405 and 1433, with intermittent gaps in voyaging caused by the deaths of the sitting emperors and the ensuing shifts in power among the Ming princes, Admiral Cheng Ho steered a large fleet on a total of seven expeditions to the "Western Sea", as the Indian Ocean was known. Cheng Ho began by visiting lands that lay along trade routes that had been well established since the Han Dynasty (206 BC–AD 220). The fleet visited, received tribute from and in some cases established direct diplomatic relations with, peoples in Java, Sumatra, Vietnam, Ceylon and other points along the coast of mainland India and throughout the Indian Ocean. His vessels were large enough and well-supplied enough to transport a pair of giraffes back to Beijing as a gift for the emperor in 1415.

---

[36]   Levathes (1994).
[37]   Needham (1991).

While in Africa Cheng Ho established relations with the Kingdom of Mogadishu and other nearby African states, passing an invitation from his emperor to send ambassadors back to the Ming throne. The nobles of "Zengdan", a Chinese name that translates as "Land of the Blacks", responded enthusiastically, and brought along even more exotic gifts to emperor Yong'le and his court. China and Africa, distant trading partners for nearly two thousand years, were now in direct contact. During his last voyage, Cheng Ho is believed to have travelled as far as modern day Iran. A Chinese map from the early 15th century illustrates southern Africa and Madagascar in some detail, and on one of his last voyages there has been speculation among scholars that Cheng Ho may have actually accomplished a rounding of the Cape of Good Hope, and sailed his ships far into the Atlantic, although there seems to be little hard evidence for this.

What complicates historical analysis of the extent Cheng Ho's explorations is the tragic fact that the logs of his last two voyages were burned by the sitting emperor once the fleet was recalled in about 1433. Cheng Ho's great patron, emperor Yong'le, died in 1424, and while Cheng Ho managed two final voyages after the death of Yong'le's immediate successor he was facing increased resistance from the ruling classes. At this time the Ming Dynasty was entering into a period of economic decline and the ruling class wanted more resources to combat the restive Mongolian tribes in the north.

At the same time, there seemed to be a philosophical sea change among "Mandarin Confucian bureaucracy" of the time, and many in power felt that the growing number of "foreign" ideas coming into China as a result of global exploration and encounters with different cultures represented a threat to the status quo, and therefore a threat to the power base of the ruling landlord class. What was contained in those last two voyage logs? What knowledge was frightening enough to the ruling class that they felt compelled to destroy them?

In a complete reversal of its previous direction, Ming China began to turn inward; the landed nobles argued that funds would be better spent at home on programs such as irrigation, granary or canal projects, and defense against the Mongols mostly through the extensive expansion of China's Great Wall. Despite the glimpses of the multiple and varied worlds beyond their own Middle Kingdom during Cheng Ho's travels, now the ruling classes decreed that they, as the Sons of Heaven, had nothing to gain or even learn from the outside world, sealing themselves behind their Great Wall. No one stood up to take Cheng Ho's place, and the magnificent fleet rotted away at the docks. Royal edicts were even put in place that forbade Chinese citizens to leave China, and by 1500 it was a crime to build a vessel capable of anything beyond coastal fishing, and officials were ordered to arrest those who still went to sea, and destroy vessels capable of seagoing journeys: "Shipyards disintegrated, sailors deserted, and shipwrights, fearing to become accomplices in the crime of seafaring" became harder and harder to find.[38]

---

[38]   Boorstin (1985): 200.

Some historians claim than instead of viewing the cessation of maritime exploration after Cheng Ho's death as a failure, the voyages ended for practical reasons, and Cheng Ho's voyages under the emperor Yong'le should be seen as expanding China's influence in the region. Indeed, Cheng Ho's expeditions did expand China's influence, and brought back a world of knowledge to the Ming Dynasty – this is not in doubt. But it was expensive. Had the Chinese continued on their course, it is quite likely that they would have "discovered" Europe – and not the other way around – and it is also quite likely they would have avoided centuries of humiliating occupation by European powers. China, the "world leader in science and technology in the early fifteenth century, was soon left at the doorstep of history."[39] Had the ruling Ming court known the outcome of being "the discovered," how might they have thought differently about the cost?

Cheng Ho's voyages were a positive development for China, but for whatever reason, they stopped. The Apollo landings brought us much knowledge as well – but they, too, stopped. In the words Apollo 11 astronaut Edwin Aldrin:

> "For one crowning moment we were creatures of the cosmic ocean . . . Yet an eerie apathy now seems to afflict the very generations who witnessed that event. The promise of a sustained, vibrant and growing human presence on the Moon has died a pathetic almost incomprehensible death."[40]

While the example of Cheng Ho's fleet is often cited by advocates of continued space exploration as metaphor for a fate to avoid, none have drawn the parallels as strikingly as Professor John Lewis:

> "The last three flight-ready Saturn 5 boosters [of the same type that took humanity to the Moon], already built and paid for, were laid out as lawn ornaments at Cape Canaveral, Marshall Space Flight Center, and the Johnson Space Center, to rust into ruin . . . . The tools and dies for making the Saturns were collected and sold for pennies a pound; a $20 billion investment in the future was melted into dross. The plans and the blueprints and operating instructions for the Apollo hardware was declared surplus. One set of Saturn 5 plans was donated to a Boy Scout paper drive. The last apparently complete set of plans was sent to the Federal Record Archives in Atlanta. My attempts to find them several years ago met with no success: the plans have apparently been lost. The fleet has been destroyed. The plans are gone. The eunuchs have won the day."[41]

Clearly, while the technologies are important, they go nowhere without human interest, imagination, or will. In the case of ancient China, and the USA

---

[39] Levathes (1994): 20.
[40] Aldrin is quoted in Shapiro (1999).
[41] Lewis (1997): 4.

more recently, interest and will were lost, with exploration-related results equal to those of the collapse of any civilization.

Nuclear Conflict

The Nuclear Age began with the detonation of the first fission device by the United States August 6th 1945; shortly thereafter, on March 1, 1954 at Bikini Atoll in the Marshall Islands, a 15 megaton device – 1000 times more powerful than the Little Boy bomb dropped on Hiroshima – was detonated, and on October 31, 1961 the Soviet Union detonated a weapon code named *Vanya*, the most powerful nuclear explosive device ever tested, yielding a force of 50 megatons, more than 1400 times the combined power of the two nuclear explosives used in World War II, (Little Boy (13–18 kilotons) and Fat Man (21 kilotons). At the height of the Cold War, in the early 1980s, there were approximately 21,000 nuclear warheads split between the United States and the Soviet Union; many additional fission weapons were in Chinese and Indian stockpiles, and several fusion devices were under control of the governments of the UK, France and presumably Israel as well.[42,43,44]

Fortunately, despite a few close calls, the world's nuclear armaments were never used after World War II. Industrialized civilizations maintained control of nuclear weaponry and had a realistic understanding of the destructive power of their arsenals. Today, however, the proliferation of radioactive weaponry has begun. The UK's National Consortium for the Study of Terrorism associate Dr Gary Ackerman has noted that while nuclear terrorism is unlikely to lead to collapse of civilization directly – in the terms we examined earlier – it could of course cause terrible destruction.[45] How many centers of civilization would have to be destroyed at a given time for civilization to collapse through certain stages is unknown, and, of course, not something anyone wants to find out. Ackerman's suggestions for preventing such disasters include centralizing nuclear material production globally, severely restricting international transfer of this material and control of 'loose' nuclear material already out of state control. While the scenario of simultaneous nuclear destruction of civilization centers is dire, there is perhaps a more realistic and terrifying possibility: that

---

[42] *Castle Bravo* was the code name given to the first US test of a dry fuel thermonuclear hydrogen bomb device, as the first test of Operation Castle, a joint venture series of high-yield nuclear tests by the Atomic Energy Commission (AEC) and the Department of Defense to test designs for an aircraft-deliverable thermonuclear weapon. Regarding the test at Bikini Atoll, over 200 islanders were evacuated, and a crewman on a Japanese fishing boat later died as a result of radiation poisoning (see O'Keefe (1983)). The Russian 50 MT H-bomb was code named "Vanya" (or "Tsar Bomba".

[43] For an overview, see DeGroot (2005).

[44] Adamsky and Smirnov (1994).

[45] See James Martin 21st Century School (2008).

nuclear detonation and/or computer hacking could convince nuclear states into believing they had been attacked, initiating a larger conflict, as in pitting the US and Russian nuclear arsenals against one another. Naturally, better controls on the current nuclear arsenals, and their entire dissolution – long a priority of Carl Sagan for precisely the reason that they pose such grave risk to humanity – are warranted.

In short, the chief dangers to human civilization today, regarding nuclear weapons, are:

- Accidental launch of nuclear exchange.
- Nuclear terrorism ('hot' or 'cold' nuclear releases).
- Restart of Cold-War hostilities followed by nuclear warfare.
- Proliferation of nuclear weapons and use by unstable states.

In the case of danger from nuclear weaponry it is unclear how that danger ranks as potentially leading to either human extinction or the collapse – in the worst case, irreversible collapse – of civilization. But, as we have pointed out, in some ways a collapse into a Dark Age would be equally as terrible as all-out human extinction. There is no guarantee that the essentially humanistic civilization that we value – despite all its flaws – would ever re-emerge. Humanity has lived in many modes for millions of years, and archaeology has shown up the fallacy that human evolution has been progressive in the sense that it has been internally driven towards the assembly of modern, humanistic, post-Enlightenment Western civilization that we value. Naturally, the point is to prevent these terrible scenarios in the first place. And this can be done at the same time as building insurance for our species in the worst-case scenario, that is, by extending the human range away from Earth alone.

Pandemics

Prokaryotic bacteria have been evolving in lockstep alongside their eukaryote cousins for hundreds of millions of years. As animals became more complex, bacteria remained simple, adaptable, quick to change and profligate in their reproduction. They remain some of the simplest forms of life on Earth, but they should be considered as potentially dangerous as nuclear weaponry. Microbes can evolve quickly, and if a deadly microbe were to spread, or to be made to spread rapidly – easily facilitated by jet travel – a global pandemic could cause serious trouble for our species. Our vulnerability to microbes is clear when we consider that we get a new flu shot every autumn because every year new mutations of the influenza virus evolve, posing renewed threats to the health of the global population. There are many examples in the past of natural epidemics involving bacteria; in Medieval Europe, of course, there was the bubonic plague.

Today, pathogenic bacteria include *Mycobacterium tuberculosis*, the culprit behind tuberculosis, *Mycobacterium leprae*, the cause of leprosy, and *Streptococcus pyogenes*, a bacterium responsible for causing many human ailments including toxic shock syndrome, necrotizing fasciitis (flesh-eating syndrome) and scarlet fever. The common "staph" infection, *Staphylococcus aureus*, has managed to mutate to the point where it is resistant to all but a few of the latest (5th generation) antibiotics, while retaining its virulence. Perhaps best known in recent years and most feared for its virulence, *Bacillus anthracis*, is the causative factor of anthrax, which has been weaponized by a number of governments, and appear on occasion to have been leaked from government labs.

Other potentially dangerous biota include viruses, which can have devastating effects on human populations. In 1918 over 500,000 died in the United States from the Spanish influenza, which killed approximately 50–100 million people worldwide. While these numbers are high, we should keep mind that even in the 1918 epidemic over 95% of the global population survived.

Whatever its effect, a global pandemic could be natural or caused by humans. In *Vectors of Death*, anthropologist Ann Ramenofsky discusses the smallpox epidemic deliberately introduced to the plains-dwelling Native Americans:

> "Among Class I agents, *Variola major* holds a unique position. Although the virus is most frequently transmitted through droplet infection, it can survive for a number of years outside human hosts in a dried state ... As a consequence, *Variola major* can be transmitted through contaminated articles such as clothing or blankets ... In the nineteenth century, the U.S. Army sent contaminated blankets to Native Americans, especially Plains groups, to control the 'Indian problem' ..."[46]

During the Second World War, the Japanese conducted experiments using plague in China and, as did the United States, conducted research into other pathogens: "Both programs studied the same gamut of organisms chief among which were plague, anthrax, glanders, typhus, dysentery and cholera. Researchers in both countries sought the same types of data, concerning the best routes of infection, the optimum particle size, the minimum infectious and lethal doses, and the ideal size, shape, density and persistence rates of pathogen clouds."[47] During the Cold War, biological weapons research continued in both the United States and the Soviet Union. Fortunately, however, biological weapons have not yet been used on a large scale.

Could a global pandemic of a deadly virus cause human extinction, or the collapse of civilization? It is not difficult to imagine tremendous population decline in the case of a virus or microbe that might elude our efforts at extermination. It is likely that we will find out, because it is impossible to know how all of the world's microbes and viruses will mutate through time. And, as with other global threats, it is not necessary to envision extinction as the only threat; if depopulation were so great that the elements of global civilization were

---

[46]   Ramenofsky (1988): 148.
[47]   Regis (1988): 148.

disintegrated, we would easily of course enter a new Dark Age, a tremendous loss for our species and for knowledge, with no guarantee that what we value of civilization today would ever reassemble.

The Worst-Case Scenario

The use of nuclear weaponry, the spread of disease, the impact of Earth by a large piece of space debris … in the long run it seems that at some point something will be capable of rapidly depopulating the major urban centers of the world, leading to a rapid disintegration of the integrated elements of modern civilization, such as water treatment and communications facilities, economic arrangements, medical support and communication systems, and so on. Although it might be possible to repair these systems in time, chaos often follows even limited disasters, and if the immediate after-effects included multiple nations striking out against one another, in an attempt at perceived retaliation, a global war could result, causing further chaos. What would this chaos mean, exactly?

We can recall the significance to civilization of *non-food production specialization*, in which many people are engaged not in food production, but specialized tasks, including data management, communications and manufacturing. For example, our technological civilization requires multitudes of engineers and other specialists trained to build, maintain and operate complex machinery, such as airliners. Even cars are not simple; most of us know how to drive one, but most of us cannot build one. We know how to plug radios into an outlet, but if there were no electricity, most of us would be stretched to try to build an electrical plant and a power grid. Consider modern medicine, which has since the 1950s revolutionized health care, and, for example, cut childbirth-related deaths of mothers by 50%. What about libraries, or other centers of knowledge, which are staffed by information specialists whose lives themselves are protected by firemen, EMT's and doctors? If enough specialists around the world were killed or incapacitated, if enough key people were unable to perform such specialized roles because the agricultural systems that feed them had broken down, the structured arrangements of matter and energy that we term 'civilization' would collapse. Urban centers would be depopulated as survivors moved out to rural areas simply to be able to grow food to eat. Contagious disease would likely spread after such a collapse; Figure 4.6 shows an Inka smallpox victim in the 16th century AD.

Consider all that knowledge that would be lost in such a disaster, and how much time might be spent to rebuild civilization. We can certainly forget, in this scenario, researchers focusing on the stars or microbes. Nearly everyone would become farmers, and, if past dark ages are a gauge for the future, not farmers in an agrarian utopia, but in a feudal system in which local lords exert their will through mercenary forces, and democratic principles are quaint.

It is important to recall that such collapse was as difficult for the members of ancient civilizations to envision as it is for us. Rome flourished for nearly a

Figure 4.6. Inka Plague Victim. Copy from a 16th Century Manuscript by Cameron M. Smith.

thousand years, but disintegrated in the space of decades due to political changes. The Maya and Angkor civilizations flourished for nearly the same period, and were disintegrated by ecological disasters that undercut their agricultural foundations. Confined to one planet, modern industrial civilization could easily suffer the same fate. Some consider this inevitable; civilizations rise, flourish, then decline and fall, only to have other civilizations replace them. But while this has occurred in the past, there is no known mechanism that ensures it happens in a systematic way – although wilfully ignoring of the lessons of the past seems to be a good way to ensure it will happen again. We feel it is worth considering these lessons, and ensuring a future for our civilization in the short term, and our species in the long term, by establishing other branches of civilizations away from Earth. In that case, collapse of one might not mean the collapse of all.

Certainly world events will necessarily reshape the political map of the Earth and, depending on a variety of circumstances, by the end of the next half-century the technological societies of Europe and North America might become technological, economic and political backwaters. Institutions like NASA, ESA and Roscosmos (the Russian Federal Space Agency), for all their past achievements, could easily be defunded (their funding cannot drop much lower and still pay for human space exploration) and dissolve if survival becomes a greater priority. Figure 4.7 is a sobering reminder of how a large space debris impact might look like on Earth; one can imagine the tremendous shock and disturbance to our relatively fragile civilizations in the face of such a cataclysmic event (and Figure 4.8 shows the location of known impact features on Earth, reminding us that it is only a matter of time before another large impact). But even in the absence of such giant upsets, simple economic or political issues could cripple technological efforts needed for space colonization which, in the long run, is the only way for humanity to survive. Still, perhaps Asia, led by

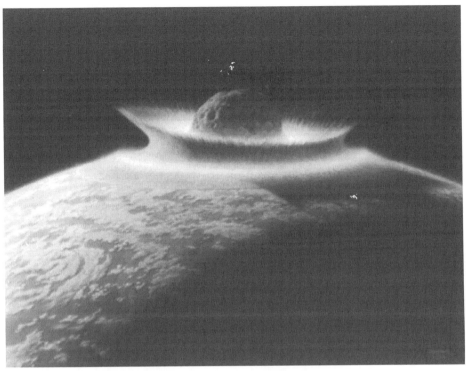

Figure 4.7. NASA Representation of a Large Comet Impact with the Earth. NASA public domain image by Don Davis.

China, will end up leading the world in education, medicine and technology. If this happens after a collapse in the West, we will be exceedingly fortunate, because another civilization will have picked up where Europe and her children in the New World left off.

The Only Way Out is Up

In the words of Astronomer Royal Martin Rees, "Long before the Sun finally licks the Earth's face clean, a teeming variety of life or its artifacts could have spread far beyond its original planet; provided that we avoid irreversible catastrophe before this process can even commence."[48] Humans have tremendous and even unknowable potential. But, as the Han Chinese said, we live in "interesting" times. Critical times, says Rees: "in the twenty first century humanity is more at risk than ever before from misapplication of science."[49] It is not only science that

---

[48]   Rees (2003): 182.
[49]   Rees (2003): 186.

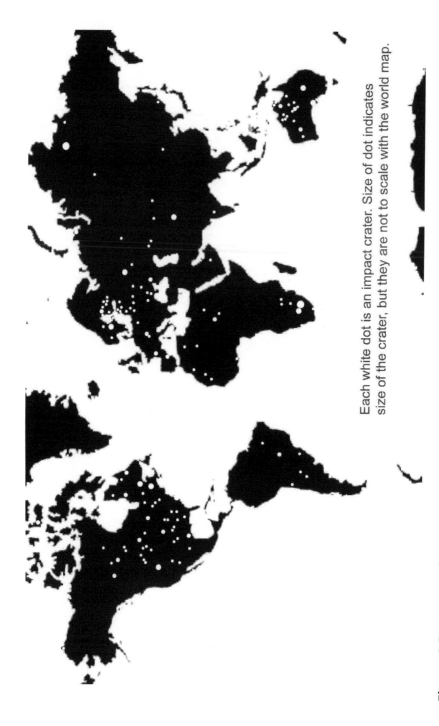

Each white dot is an impact crater. Size of dot indicates size of the crater, but they are not to scale with the world map.

Figure 4.8. Location of Known Impact Craters on Earth. Note that, being 70% covered with water, many impact sites on Earth are unknown, and that even on land, erosion and vegetation growth over long periods have probably made many other impact features difficult to find. Image by Cameron M. Smith after Grieve (1994).

could be misapplied; an endless-growth model of commerce is of course impossible, but that does not prevent us from pursuing it. Many fisheries today are near collapse, after just a few decades of industrial fishing, and every fishery ever discovered has been chronically over-fished.[50]

In an interview with a Canadian television station in 1993, Carl Sagan expressed his hope that space exploration and eventual colonization would occur, as the costs continued to decline, and the urgency became clearer to people.[51] In this chapter we have attempted to impart that certain urgency, though without hysteria. We have shown examples of natural calamities that could occur on Earth, and speculated on easily-imaginable human-made catastrophes, catastrophes that – in each case – could either bring about the downfall of global civilization or altogether extinguish the human species.

For all of these reasons, as hard to imagine as they may be, we must begin to develop outposts of humanity outside of Earth, if we are genuinely concerned about our collective future. We buy insurance plans against our individual lives to protect our families. We should do the same, by space colonization, for our offspring. Plenty of others have given similar warnings, but today we write for a new generation.

The future is simply the result of daily decisions, and if humanity is to emigrate from the home planet, to expand geographically like any mature species, some people of the next generation will have to make similar decisions to those of the 'crazy dreamers' like Burt Rutan, the aircraft designer who build the world's first privately funded and piloted spacecraft; Richard Branson, who has established space tourism; Franklin Chang-Diaz, designer of the Variable Specific Impulse Magnetoplasma Rocket (VASIMR) (an engine claimed to have the potential to cut flight time to Mars to 40 days); or Dr Dava Newman, an MIT professor who designs advanced life-support garments for the human exploration of Mars (see Figure 8.3). None of these developments can be said to address the immediate and legitimate concerns of humanity, but in the same way we do not quit creating art, or give up on intangibles when things are materially difficult; indeed it could be argued that in such times it is most important to retain our dreams.

Responsible parents want success for their children; they want them to survive and flourish, even at one's own expense. The risks people take and the strides they make today will determine where our descendants stand in the future. It makes sense for us to do all in our capacity to ensure that there will be abodes of humanity off of Earth, to ensure the survival our species, to continue life. It makes fundamental sense to us to continue to expand our understanding of ourselves and of our Universe, to continue grow in knowledge and wisdom, in short, to continue to evolve.

---

[50] Jackson et al. (2001).

[51] See interview with Carl Sagan, 'On Mars and the Excluded Middle' at: http://www.youtube.com/watch?v=iCZtLVim94Q&feature=related

5 False Choices: Common Objections to Human Space Colonization

"The idea of [space] migration is really only taken seriously by demented dreamers and sci-fi buffs who conveniently ignore the fact that the Universe is rather large."

Gerard DeGroot, 2010[1]

" ... space [colonization] (1) is too hard,(2) takes too much energy, (3) is too dangerous and (4) involves distances that are too enormous ... Most people are too lazy, too poor, and too stupid to even think about stepping up to a challenge that huge."

Science blogger Al Fin, 2008[2]

"Overall, the Universe is not a very inhabitable place for humans. And so the problems of the Earth must be solved on Earth."

Sir Martin Rees, Astronomer Royal, 2010[3]

Introduction

We have seen how humanity's adaptive trick of niche construction has allowed our relatively frail genus, *Homo*, to proliferate across the globe. We have also argued that this survival technique is a normal part of our evolution, and that to cease it now would be to move counter to evolution itself, a poor choice. Before describing some ways to better inform our adaptation to space it is necessary to dismantle some of the more common critiques of the concept of human space colonization. Inasmuch as wide public support will be necessary for human space colonization – and we argue that while that would be useful, it is decreasingly necessary as space is increasingly becoming privately accessible – then the usual

[1]   DeGroot (2006): 265.
[2]   The quotation is from blog entry *A Skeptical Look at Space Colonisation* (14 April 2008) available online at http://alfin2100.blogspot.com/2008/04/skeptical-look-at-space-colonisation.html.
[3]   Rees is quoted in an interview ("There is Always Room for Mysteries") with *The European* magazine, available online at http://www.theeuropean-magazine.com/31-rees-sir-martin/32-cosmology.

arguments against human space colonization will need to be defused. Our central point in this chapter is that while there are plenty of Earth issues to address, human space colonization would not be about abandoning Earth, but about adapting to space; two very different things.

Also, we argue that human space colonization should be valued and conceived as a beginning, not an end, and that in the end, only space colonization will save our species from the fate of 99% of all Earth life since the beginning of the fossil record: extinction. To encourage that perspective, and a culture to carry it, in this chapter we systematically dismantle the usual critiques of the project of human space colonization. Such critiques normally fall into one of two broad categories; that space colonization either *cannot* be done, or *should not* be done.

The Usual Objections: Humans *Cannot* Colonize Space

Space Colonization Would Be Too Technically Challenging

Harvard astrophysicist Eric Chaisson, of the Wright Center for Science Education, writes off space colonization as technically simply 'over the top':

> "Now that the entire planet seems headed toward a worldwide glut of people, why can't modern spacefarers just track down some stars having nice new planetary abodes for Earthlings to emigrate and settle? Though advocated by some, this approach is nonsense. Interstellar expansion is acutely more difficult than most people realize. Despite the rash of recent findings of exoplanets around stars beyond our Sun ... astronomers are presently unaware of any other Earthlike planets. Even if we did know of such a star system with homelike abodes, and even if that system were among the closest to us, it would still be far too distant to explore with our current skills. The idea of trucking hordes of people toward other planets to relieve crowding on Earth is another problem altogether. The most advanced nations now have trouble keeping just a few astronauts in space for a few months at a time. Interstellar emigration is surely out of the question now, and studies of futuristic spaceflight techniques based on the known laws of physics suggest that it might never become feasible."[4]

In this passage, Chaisson treats the subject matter coarsely, raising a straw man: that space colonization advocates are looking for a 'quick fix' to immediately begin transporting humanity to distant stars. Actually, all serious thinkers about human space colonization have proposed a slow, incremental,

---

[4]   See comments online at: http://www.tufts.edu/as/wright_center/cosmic_evolution/docs/fr_1/fr_1_future1.html.

conservative approach, beginning with initial colonies on nearer bodies such as the Moon and Mars, and expansion out from there following decades of Solar System reconnaissance by space probes. And while the world's space programs have to date been carried out for a wide variety of reasons, none has been unduly brash:

- First, humanity sent out from Earth low-orbit satellites.
- Second, humanity sent up orbiting individual people.
- Third, humanity sent up pairs of trios of people into Earth orbit.
- Fourth, humanity sent small crews to the Moon.
- Fifth, humanity has for the last 30 years sent larger groups of people into Earth orbit, learning an immense amount about sustaining human life in that environment.

Chaisson also overstates the difficulties of keeping people alive in space so far: while this has been difficult and costly, that has not been because it is essentially difficult or expensive, but largely because humans-in-space activities so far have been largely been designed not for optimal engineering or low cost, but more often to satisfy space industry contractors; in the case of NASA, critical design elements of the space shuttle were strongly influenced by the need to keep taxpayers happy, which translates into employing contractors, nationwide, even if this complicated (and even compromised the safety of) the space shuttle. It also ignores the point that NASA and other space-access agencies have deliberately built up an exclusivity, a 'Right Stuff' mythos about space access, in part to maintain their own image and funding.[5] Finally, Chaisson chooses the perspective that, in the end, interstellar propulsion 'might' not be feasible; this adds nothing to the argument because, just as equally, it might indeed be feasible. Ultimately, Chaisson's critique (a) overstates difficulties by not addressing their real source, (b) sets up a false image of what serious space colonization advocates are proposing, and (c) chooses pessimism rather than optimism. Regarding this last point, we should recall that sixty years ago, spaceflight itself was only a theoretical possibility. Today, the Voyager probes, launched in the 1970s, continue out past the edge of the Solar System, and many of the Solar System's moons and planets have been extensively explored by probes. In the last two decades there has not been a moment at which at least one human being was living not on Earth, but in its orbit.

Another space scientist, Jeffrey Bell, is frustrated by modern space colonization advocates' apparent lack of appreciation for the technical hurdles involved in space colonization. In a 2008 *space.com* article titled "The Dream Palace of the Space Cadets", Bell wrote that:

> "Even worse, [space colonization advocates] have no idea how much space travel costs, or how these costs compare to other areas of human

---

5   See Vaughan (1996).

activity like war or mountain-climbing. They think that Will is all you need to colonize the Solar System – they have no concept of the political, financial, and technological investment that it would take. But the small fraction of the pro-space community I meet in person seems tame compared to the internet space chat community. One regularly finds long discussion threads on politically impossible ideas like a one-way Mars suicide mission, financially impossible ideas like building spaceships on the Moon, and technically impossible ideas like ion-powered space blimps. In all these discussions, the few informed people who try to point out the massive problems with these ideas are swamped by a much larger number of enthusiasts who clearly don't know enough basic science or engineering to even understand the issues. I get even more frustrated when I visit the web sites of the various space advocacy groups. They are a pale shadow of the L-5 Society and the Space Studies Institute (both of which I joined in the 1970s). Many of these organizations seem to live in a dream palace of their own creation that has no relationship to the real world at all."[6]

There are several problems with this critique. Principally, while it's true that the technical issues will have to be addressed, they will not be addressed if the project is never imagined in the first place. Rocket pioneer Robert Goddard was not born knowing how he would get to Mars, but his imagination suggested it was possible, and he designed rockets that could make it possible; and indeed, in the end, made it possible. Of course, Goddard was rare in his combination of imagination and technical training, and the point is this: not everyone needs to know the technical details. Everyone uses cell phones, but how many know their technical details? Regarding the lack of technical detail in current space-colonization societies' plans, that is a temporary matter; if the will can be raised, the argument is, those details will then be formulated. And, just because this is the current state of affairs, we argue it is better to improve those organizations rather than abandon them and their goals.

Ultimately, it seems that Bell is undervaluing the significance of imagination. While the technology for space colonization is necessary to make it happen, it is not sufficient. It will require both imagination and technology, and the technology will probably not be forthcoming unless its use is imagined first.

Examples of the transformation of dream to reality from our experience on Earth are many. One is seen in the case of the French engineer Thome de Gamond. In 1833, he began to investigate the geology and hydrography of the English Channel with an eye to designing a tunnel to link England and France. After years of preparatory study, in 1855 de Gamond took a boat offshore, took a breath and "dived to the seabed, his feet weighted with 160lb bags of flint and his ears plugged with lard pads ... He carried 10 inflated pig bladders on his waist, which

---

[6]   Bell (2005).

helped him to surface when he had finished his work and cut free the bags of flint."[7] Naturally, de Gamond's scheme was considered by various engineers and critics as mad, pointless, a technocratic fantasy, too expensive, and so on. Despite these critiques (and eel attacks in these 30m (100-foot) free dives), de Gamond learned about the bottom conditions and in the following years proposed no less than eight tunnel, tube and bridge designs to connect France and Britain. While his work was largely ignored early on, by 1868 the first Channel Tunnel Committee was formed in London, resulting in several real-world tests of tunneling technology. Between 1984 and 1994, the channel tunnel was built:

> "So why is it that the Channel Tunnel is almost complete when everything was stacked against it? ... while the directors, politicians, and bankers were attacking each other, making headlines and predicting the project's imminent collapse, the construction workers were simply getting on and building it."[8]

In the same way, the will and imagination to make something work has often been as important as the hardware itself. Discussing his building of the original rocket engines for the *Bell X1*, the first craft to fly faster than the speed of sound in the late 1940s, Rocket Installations Engineer Wendell Moore stated:

> "We had valves in that plane that were boilerhouse valves, old brake valves – anything we could fix up in a hurry. To save time, we'd use standard castings and then design our special fittings to match the size. When we got through, we had a big fantastic piece of junk that you'd suspect might belong on a tractor. But we made it work."[9]

The point is that, ultimately, the hardware of space colonization will not assemble itself, and the roles of imagination and will should not be under-emphasized.

Having said all of this, one way to begin to address the actual challenges of human space colonization is to consider the task more specifically than simply discussing, for example, 'space'. There are many space environments, and they differ as do Earth environments, as discussed in Peter Eckart's *The Extraterrestrial Environment*. This delineation and characterization of space environments is indistinguishable from the first step of Earth-based ecological studies that begin with the delineation of components of an environment (and, normally, their interactions). From an evolutionary and adaptive perspective, in this case the phenotype is the spacecraft (and its occupants) and the selective environment is composed of the various conditions in various space environments. So, rather than space being taken as a 'black box', we can begin to parcel it out into its constituents, allowing us to better understand just what will be needed to adapt

---

[7] Varley (1994).

[8] Anderson and Roskrow (1994): xiii.

[9] Mallan (1955): 126.

to them technically, biologically and culturally. This is a far cry, as it should be, from simply saying that it will be too hard to 'adapt to space'. Eckart's book is organized as an extended discussion of five main adaptive challenges of space; we present these below, but will revisit them later in this book:[10]

- Radiation in free space.
- Gravity.
- Vacuum.
- Magnetic fields.
- Local planetary environments (e.g. Mars and the Earth's Moon).

Each of these variables differs in different parts of 'space', and some make some areas essentially uninhabitable while others remain habitable.

In the same way as Eckart, space scientist Alan C. Tribble's *The Space Environment: Implications for Spacecraft Design* was written to even more sharply focus on the variable conditions of space environments, in part as a response to the loss of space shuttle *Columbia* in 2002. Tribble contends that closer consideration of the space environment that he is concerned with – Low Earth Orbit – could make future space flight safer. Tribble's delineation of Low Earth Orbit environments includes several hazards and design mitigations for these hazards, as summarized in Table 5.1.[11] (Note that for each environmental variable shown in the table there are design guidelines that can be adopted to reduce risks, showing that 'space colonization' is not as simple as 'it can be done' or 'can't be done'. Table 5.1 is adapted from Tables 2.12, 3.11, 4.7 and 5.9 of Tribble (2003); while its contents will be familiar to space engineers, the point is that for many people, partitioning 'space' into these elements could in fact make human adaptation and colonization of space more conceivable through disambiguation.

In a similar way, we should be aware that there exists a spectrum of readiness for a given technology, also something not always appreciated by critics of the technical issues of human space colonization. Peter Eckart, mentioned above, handily lists the following eight 'Readiness Levels' of this spectrum, something to consider when we hear the critique that space colonization technology 'isn't ready':

- Readiness Level 1:   Basic principles observed and reported.
- Readiness Level 2:   Conceptual design formulated.
- Readiness Level 3:   Conceptual design tested by analysis and experiment.
- Readiness Level 4:   Critical functions demonstrated.
- Readiness Level 5:   Components tested in relevant environment.
- Readiness Level 6:   Prototype tested in relevant environment.
- Readiness Level 7:   Validation model tested in relevant environment.
- Readiness Level 8:   Design qualified for flight (use).

---

[10]   Eckart (1994).
[11]   Tribble (2003).

Table 5.1. Some Low Earth Orbit Environment Design Issues.

Low Earth Orbit Environmental Variable	Design Element	Design Guidelines
The Vacuum Environment	1. Materials Selection	1. Choose UV resistant and low-outgassing materials and coatings
	2. Configuration	2. Vent outgassed material away from sensitive surfaces
	3. Margin	3. Allow for degradation in thermal/optical properties on orbit
	4. Materials	4. Consider vacuum bakeout of materials before installation in vehicle
	5. Pre-Treatment	5. Provide time for on orbit bakeout during early operations; provide cryogenic surfaces for the opportunity to warm up and outgas contaminant films
The Neutral Environment	1. Materials Selection	1. Choose the materials that (a) are resistant to atmospheric oxygen, (b) do not glow brightly (if optical instruments are present), and (c) have high sputtering thresholds
	2. Configuration	2. Aerodynamic drag may be minimized by flying the vehicle with a low cross-sectional area perpendicular to ram; orient sensitive surfaces and optical sensors away from ram
	3. Coatings	3. Consider protective coatings for surfaces that are susceptible
	4. Operations	4. If possible, fly at altitudes that minimize interactions
The Radiation Environment	1. Uniform Surface Conductivity	1. Make exterior surfaces of uniform conductivity if possible
	2. Electrostatic Discharge Immunity	2. Utilize uniform spacecraft ground, electromagnetic shielding, and filtering on all electronic boxes
	3. Active Current Balance	3. Consider flying a plasma contactor or a plasma thruster
The Plasma Environment	1. Shielding	1. Place structures between sensitive electronics and the environment in order to minimize dose and dose rate
	2. Design/Parts Selection	2. Utilize parts that have sufficient total dose safety factors and are latchup and upset resistant
	3. Redundancy	3. Oversize solar arrays and design electronic systems with backup circuitry/parts
	4. Software Recovery Algorithms	4. Install software capable of recovering the system from latchup or upsets

As social scientists, we are not qualified to evaluate the technical details involved in human space colonization – but that is not our goal. The argument that human space colonization would be too technically difficult seems to be rather short-sighted and leans more towards pessimism than optimism; and, it sometimes it can be dismantled – at least in part – by more finely considering what 'space' is. *What*, one should ask when encountering this critique, can't be done?

So while there are plenty of engineering challenges to overcome to make a success of human space colonization, it does not seem that writing off the entire project is warranted; partly because we are still learning about and understanding space environments, which helps to define the kinds of technologies we will need to adapt to them. And those technologies do not exist in an 'either/or' Universe, but one in which a 'crazy dream' today can slowly be built into an operational system – such as the airplane, for instance. Finally, it seems certain that further space exploration will lead to a better understanding of our Universe and what we need to do to live in it. In an interview about distant space technologies in the mid-1950s electronics scientist George H. Stine stated:

> "Sitting where we are [on Earth] we have a very distorted notion of physics, of the Universe. We've got this milky, murky atmosphere about us and we're always in one gravity field – we can't really visualize the ultimate possibilities of spaceflight. We have to get out there first, to know what we can do."[12]

Ultimately, the argument that space colonization would be too technically difficult to achieve fails – or is at least not as strong as it could be – for several reasons:

- Not all potential off-Earth environments for colonization have been sufficiently characterized.
- Technological leaps are common in industrial civilization (e.g. explosive growth of personal computers and communications devices that make life today significantly different from just a decade ago), such that ruling certain advances out might be over-pessimistic.
- Good, realistic plans have been forwarded, and have not yet been comprehensibly falsified (and the only reasonable way to test them is to try them out).
- Technical advances specific to space colonization will not arise by themselves, but will require the motivating forces of will and imagination.

---

[12] Stine is quoted in Mallan (1955): 315.

Humanity is Not Meant to Live in Space

A common argument against space exploration and colonization is that humanity is not 'meant' to live in space. Gerard DeGroot recently made this argument in his evaluation of the Moon exploration missions:

> "The Moon voyage was the ultimate ego trip. Hubris took America to the Moon, a barren, soulless place where humans do not belong and cannot flourish. If the voyage had any positive benefit at all, it has reminded us that everything that is good resides on Earth."[13]

This is a poor and dissolute argument on many levels. First, biology recognizes no purpose to or in evolution, so humanity is neither *meant to* do anything, nor is humanity meant to *not* to do anything; if there is meaning in evolution, G.G. Simpson pointed out, it is supplied by humanity itself. Second, saying that the Moon is a barren and soulless place is simply an opinion; the same is said of the high Arctic, or the deserts of Australia (for example), both areas of the world where entirely modern humans have thrived for thousands of years; if DeGroot finds no soul in the Moon, that is of course his issue, not the Moon's. Third, DeGroot suggests that all that is good resides on the Earth, ironically a statement of great hubris considering that we know so little about the Universe in which we live! Finally, DeGroot questions whether the Moon voyages had a value at all, a gross oversimplification of a complex issue, as we explore throughout this book. Naturally, there were negative and positive results from the Moon exploration missions. Ultimately, DeGroot's argument is largely a reflection of his own emotions, rather than a rational examination of the issue. It ignores the fact that many humans have flourished in areas that others would consider barren wastes, and it brings up a common fallacy, that humanity is *meant* to do, or not do, something. All of these arguments are easily discarded.

A more measured examination of the challenges of human space colonization shows that the issues are more complex and subtle than can be answered by "we can" or "we can't" regarding space colonization. As it turns out, in some places and times we can, and in others, we probably can't. We review the issues of human adaptation to space hazards later in this book, but below we tabulate the chief issues involved, as identified in the last half century or so of human space flight. We should keep in mind, also, that these are only issues involved in life in Earth-orbital environments, and there are other places in the Solar System that might be better places to settle. For the moment, however, the following are the main health issues so far known, issues that can specifically be brought up to counter the broad-brush statement that 'humanity can't live in space'.

Astronauts spending more than a few days in space have so far been observed to return to Earth somewhat physically weakened. This is largely a result of the

---

[13]   DeGroot (2006): 269.

lack of the 1g gravity field of the Earth's surface 'pulling down' on the human frame, causing the body to react – with the use of muscles and bones – to maintain its shape.[14] To mitigate this issue, astronauts currently exercise both for cardiovascular and resistance health, and take supplementary nutrients. Bone loss, through calcium passed in the urine, is also related to the lack of use of the bones to support the body in the weightless conditions of orbit, and this is also mitigated with supplements and exercise.[15] Space motion sickness, effecting most astronauts within a few hours of exposure to weightlessness, is also mitigated by pharmaceuticals, and though they are not very effective, the condition's nausea and disorientation normally vanish within a few days.[16] The issue of blood cell reshaping in weightless conditions, with possible long-range effects on human health, is poorly understood, but not currently considered a 'show-stopper' that should prevent human space exploration, even in the long term, or space colonization.[17] Radiation, the most sensational issue for the general public, is serious, but a number of ways to avoid radiation hazards have been devised and they are the focus of intense investigation. A 2008 report published by the US National Research Council explored the topic thoroughly, indicating how much has yet to be learned, and how carefully the issue is being divided into research domains for the variety of space radiation issues:

> "... no single scientific or engineering discipline can provide the expertise to solve these [space radiation hazard issues] optimally ... crosscutting advice needs to come from cross-disciplinary groups of experts representing diverse scientific fields rather than from the traditional single-discipline advisory committees."[18]

Solutions to radiation hazards include shielding spacecraft and colonies from radiation, limiting exposure of people to radiation by careful scheduling and planning of spaceflights, and others we mention later in this book, including the fascinating prospect of DNA repair following radiation damage. The point is that while some radiation hazards make certain areas of space particularly dangerous, they do not, at present, represent a hazard so significant as to block human space exploration. Figure 5.1 shows a number of off-Earth environments, each with its own radiation hazards, and some technological, biological and cultural proactions that could be used to offset them. As we mentioned earlier in this book, we are not engineers; our goal here is simply to show that off-Earth environments can each be understood in their own terms, and solutions devised to address each one. Shown in the diagram are the solar particles, high-speed protons and electrons trapped within the boundaries of the Van Allen belt, and

---

[14] Barratt (2008).
[15] Task Group on Life Sciences (1988): 87–92.
[16] Japan Aerospace Exploration Agency (2003).
[17] Japan Aerospace Exploration Agency (2003).
[18] National Research Council (2008): 12.

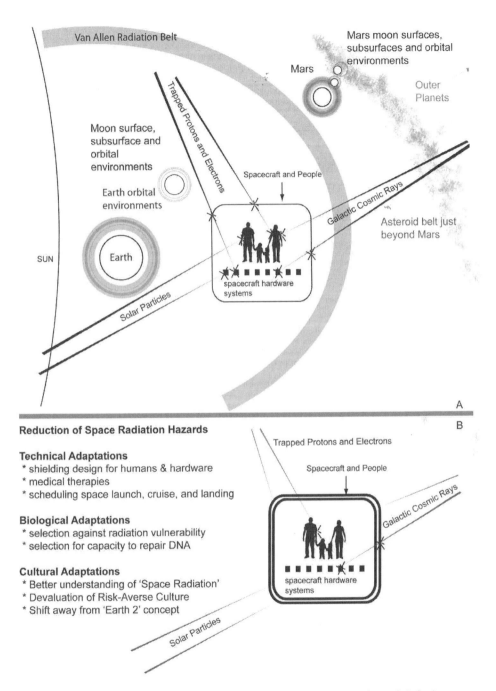

Figure 5.1. Some Off-Earth Environments, Radiation Hazards and Solutions. Image by Cameron M. Smith.

the galactic cosmic rays that pose significant hazards off of our home planet. They can pass through spacecraft, damaging hardware and the humans inside. But these hazards can be mitigated by technical, biological and cultural adaptations. Technical solutions include shielding, medical therapies to repair DNA damage, and scheduling space activity to avoid episodic dangers, such as solar storms. Biological solutions include selecting space crews and, eventually, colonists, who are less vulnerable to radiation hazards than others (simply as a function of biological variability), and whose capacity to repair DNA with the body's DNA-repair enzymes. Naturally, the ethical issues regarding such selection are large, but are not our focus in this book and will be addressed in other publications. Finally, cultural behavior can be altered to cope with radiation hazards. This could include an increased tolerance for risk – recall that, in Chapter 6, we saw that an exploratory, risk-taking ethos was important in Polynesian traditional culture – and a shift away from the idea that wherever we colonize off of Earth, we must replicate Earth conditions precisely, in an 'Earth II' mentality. No matter what we do, as we see in the lower panel, humans will still be affected by radiation, as we are even on Earth; airliner crews who fly over the Arctic already are monitored for increased radiation doses, and they are scheduled for retirement from that route at a certain point. The point is to reduce the radiation risk to something that will be biologically and culturally tolerable, and what that level is may well be different than what we consider tolerable on Earth, today.

Finally, another issue loaded with potential drama – and the subject of plenty of science fiction – is that of the effect of living 'in space' on the human psyche. Currently the selection of humans for space travel, and their work and life in space, are all carefully controlled, and while on occasion negative behavioral issues have arisen,[19] none has yet caused disaster in space and it has not yet been seriously proposed that social and psychological (not to mention cultural) issues should preclude human adaptation to off-Earth environments. These issues are summarized in Table 5.2 (based on material referenced elsewhere in this chapter) for easy reference–note that these are summaries of studies of astronauts in zero-g or low-g environments, with only a small amount of information from people living on the surface of another celestial body (the Moon) and none in simulated c.1g environments, as would be used in orbiting colonies, or other radiation and gravity environments, for example, on Mars. Note also that we identify whether space effects on the human body are *reversible* – (in other words, whether they go away on return to Earth) and whether the condition occurring in space is *heritable* – (in other words, whether it could directly affect the children of people exposed to these conditions). This is an evolutionarily significant issue revisited elsewhere in this book.)

---

[19]   See examples in Lininger (2000).

Table 5.2. Overview of Health Hazards to Humans in Low Earth Orbit.

Hazard	Observed Effects	Reversible on Return to Earth? Heritable?	Current Mitigations
Cardiovascular Weakening	Blood and other fluids migrate away from lower body in space flight	Yes No	Astronauts exercise up to two hours per day
Bone Loss	Bone demineralization on order of 3% after 10 days in space	Yes No	As above & bone (nitrogen and calcium) supplements
Muscle Loss	Muscle atrophy, particularly in lower body	Yes No	As above
Space Motion Sickness	Most astronauts experience dizziness and nausea for several days after first entering zero gravity	Yes No	Motion sickness pills
Blood System Anemia	Without Earth gravity, blood cells take on a different shape that reduces effectiveness	Yes Possibly, but unknown	None
Radiation Issues	DNA damage and interference with protein synthesis and other molecular processes	N/A: reversibility may occur in space or on Earth in the form or natural DNA repair mechanisms Yes, if damage is to sex cells (sperm or egg)	Shielding; monitoring astronaut lifetime dosage; scheduling spaceflight operations
Behavioral Issues	Stressful task environments and crowded quarters can lead to outbursts or noncooperation of space travelers	Yes Not biologically, but space travel culture could be effected by spread of negative interactions	Few formally-enacted solutions

Reviewing this table, we should keep in mind that health hazards will not have to be identified simply for humans, but also for all of the domesticated species – plants and animals, essentially – that humanity will take away from Earth; and that is not all, because each species (human or domesticate) itself coevolves with other species. For example, in humans, dozens of species of microbe, some of them important to our survival, live in and on our bodies. For these reasons, some biologists now speak not in terms of the evolution of individual organisms, but of the *holobiont*, the organism of primary interest *and all of its associated organisms.*

Finally, we must remember that none of this information addresses larger-scale issues such as cultural and linguistic change over time, long-term genetic issues, or the anthropology of colonial cultures; in part this is because humans have not spent much time in space, but also because no real human communities have ever occupied an off-Earth environment. Again, we return to these fascinating issues later in this book, where we will also consider genetic issues such as the minimum viable population and the founder effect, demography, disease and other issues to consider in serious and realistic plans for space colonization.

The argument that humanity was not 'meant' to live off the Earth has no basis in evolutionary reality. Similarly, the argument that humanity cannot live off the Earth paints with too broad a brush; in some places we can, and in others we cannot, and where these places are will of course change as our adaptive means – technology, culture and biology – all change through time. Yes, it seems clear that there are places where humanity can live off of the Earth.

Certainly life off of Earth – to which humanity is of course well-adapted – would be different than it on Earth, and certainly it would involve risks and a fear of those risks. Some of this fear is perhaps at the root of the idea that humanity simply could not live anywhere but on its home planet. But fear is a poor motivator, and while this fear will prevent some from undertaking space colonization it is likely that many would go regardless. In his novel *The Martian Chronicles*, Ray Bradbury commented on the pervasiveness of fear in human culture:

> "... there was always a minority afraid of something, and a great majority afraid of the dark, afraid of the future, afraid of the past, afraid of the present, afraid of themselves and shadows of themselves."[20]

This is certainly the case, but even German astronomer Johannes Kepler (1571–1630) recognized that there would always be someone willing to explore the Universe despite the dangers, writing (in a 1593 letter to Galileo):

> "Provide ships or sails adapted to the heavenly breezes, and there will be some who will not fear even that void ..."[21]

Ultimately, the argument that is simply not 'meant to live in space' (however phrased) fails because:

---

[20] Bradbury (1997): 163–164.
[21] Knapp (1989): 81.

- There is no intent to evolution that assigns habitats to any species.
- Species regularly adapt to new habitats.
- Species regularly migrate.
- Humanity has so far been capable of surviving in various non-Earth environments by using biological, technological and cultural adaptations; in principle these should also work for populations entirely independent of Earth, though this has yet to be tested.

The Usual Objections: Humans *Should* Not Colonize Space

Wrestling with the philosophical implications of space exploration, in 1970 Norman Mailer asked:

> "Was the voyage of Apollo 11 the noblest expression of a technological age, or the best evidence of its utter insanity?"[22]

Ignoring for the moment that this question offers a false choice between two opposites – things are almost never so simple – it is clear that most objections to the concept of human space colonization are similarly engaged with philosophical rather than technical issues. In our experience, the most common is that space colonization would divert resources better used to solve Earthly problems.

Space Colonization Would Cost Money That Should be Spent on Earth Problems

Recently the London Times Magazine carried an article that noted:

> "For a child of the Seventies like me, spoon-fed on Apollo Moon landings . . . Glorious pictures from distant worlds are all very well, but after a while they begin to feel like brochures for luxury holidays the West can no longer afford."[23]

This sentiment, that humanity can no longer afford space exploration (not to mention colonization) is at the heart of many arguments against space colonization. It is easily dismantled, however.

First, it is clear that while space colonization would be expensive in the short term, so are, nearly, all worthwhile things: wedding rings, health insurance, personal computers, cars, boats, and so on are all large initial expenses, but we take on the initial cost because we see the larger value. A good example is the US highway system, built at a cost of over $400 million 1952–1957 dollars (over three billion 2011 dollars). What space colonization would cost, in dollar terms,

---

22   Mailer (1970): 382.
23   Miller (2011): 58.

is so difficult to predict that it is not worth discussing here, but the figure would of course be high. However, the cost may be averaged over time as a worthy investment, and in this way one can reasonably conclude that, considering the threats to humanity and civilization outlined earlier in this book, the large initial outlay for a project such as human space colonization would be nothing less than an insurance policy for the genus *Homo* itself. In fact, space colonization should be considered not an end, but a *beginning*; what we are buying therefore is not an object but an idea. Finally, one may also ask what would be the cost of *not* making this investment in the human future.

Second, the costs of human space exploration so far in world history have been very low. Today the NASA budget is less than 1% of US federal spending, and even at the height of the Moon project it still consumed less than 5% of the total federal budget (see Figure 5.2). The great consumer of the federal budget today is, of course, the Department of Defense, which has been a frustration to citizens for several decades. A 2011 US Government Accounting Office report found over $135 billion in Pentagon cost overruns alone;[24] and even a decade ago Defense Secretary Donald Rumsfeld reported that nearly two *trillion* dollars of Pentagon money could not be accounted for.[25] The lesson is that if we want to find government funding for space exploration, there are plenty of ways to get it – first, for example, by eliminating such waste, much of it spent, of course on projects that in the long run only threaten rather than defend human civilization. This is an old issue, addressed in 1977 by science-fiction author and space colonization advocate Ray Bradbury:

> "As I write this, our government . . . is dishing out at a lunatic pace 300 million dollars a day to finance new tanks, nuclear ships, and other exotic weaponries planned for the Day of Super Demise, when we will not only kill but simultaneously bury ourselves, with not even birds to sing us to our rest."[26]

Since that time, little has changed: Figure 5.3 shows that nearly 60% of the 2011 US Federal Discretionary expenditures are military.

Another significant problem with the issue of cost is that it assumes that the basic human problems of Earth can be solved by an infusion of money. These basic issues have been identified many times over by United Nations studies as:[27]

- Overpopulation.
- Lack of access to education.
- Lack of access to fresh water.
- Lack of access to food.

---

[24]   General Accounting Office (2011).
[25]   Sirgany (2009).
[26]   Bradbury (1977): 9.
[27]   For example see United Nations (2005).

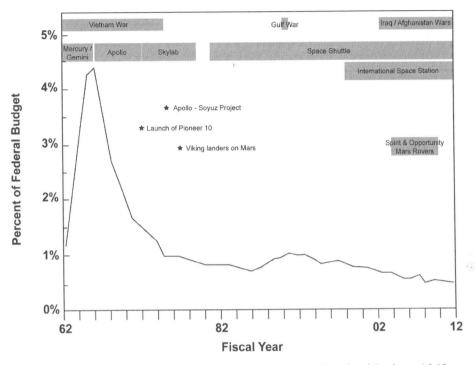

Figure 5.2. Levels of NASA Funding as Percentage of Federal Budget, 1962–2012. Image by Cameron M. Smith based on data from WhiteHouse.gov.

Regarding the first point, it is widely recognized that population control requires cultural changes more than financial solutions, and that the material elements of the issue – contraceptives, for example – are cheap and easy to distribute. A report by Netherlands' non-profit World Population Foundation recently found that "lack of knowledge of and access to family planning services are by far the most important factors determining family size" in the fastest-growing population areas, where "... if women had only their desired number of children, the birthrate would fall by 27 percent in Africa, 33 percent in Asia and by 35 percent in Latin America." [28] The UK's Galton Institute has recently noted that "lack of knowledge of and access to family planning services are by far the most important factors determining family size".[29] Similarly, even modest investment in basic education for the world's poor have significant effects, with the UN reporting recently that:

---

[28]   The quotation is from S. Bavelaar, past Executive President of the World Population Foundation, quoted in Galton Institute (2005).

[29]   See Galton Institute (2005).

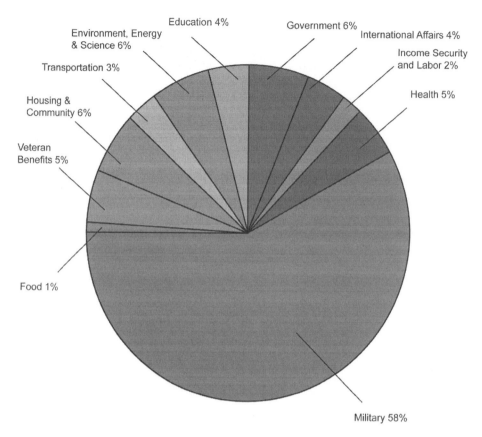

Figure 5.3. US Federal Discretionary Spending, 2011. Image by Cameron M. Smith based on data from WhiteHouse.gov.

"The abolition of school fees at primary school level has led to a surge in enrolment in a number of countries. In Tanzania, the enrolment ratio had doubled to 99.6 per cent by 2008, compared to 1999 rates. In Ethiopia, net enrolment was 79 per cent in 2008, an increase of 95 per cent since 2000."[30]

The final points, regarding water and food shortages, are also non-starters, as it is well-known that most famines are politically sponsored rather than the outcome of natural disasters; and even in the case of natural disasters, there are ample food resources worldwide that can be used to feed even the current population, if only the political issues of moving these resources to needy

---

[30] United Nations Millennium Development Goals (2010).

populations could be worked out.[31] Regarding these political issues, it should be pointed out that they, also, are unlikely to be solved by money.

Another falsity in the 'too-expensive' critique of human space colonization is that such an enterprise would consume up all available funds; but, of course, humanity can do more than one thing at a time – and in fact, waiting until the funds are prepared might be penny-wise but pound foolish, if we are unlucky and plague or other disaster intervenes before humanity spreads into space.

Yet another falsehood is that space exploration and colonization have to be as expensive as they have been to date. As mentioned earlier, NASA projects are not always designed to minimize cost, but often to maximize taxpayer involvement by awarding contract work. In *The Case For Mars*, space colonization advocate Robert Zubrin pointed out that an early NASA plan for human exploration of Mars was rejected outright by congress because of its $400 billion price tag, a cost Zubrin claimed (with some convincing numbers) that he could reduce substantially simply by being outside of the NASA-contractor loop.[32]

This relates to the other, final and important point regarding costs: that as space access shifts away from government and towards private enterprise and individuals, the need to extract taxpayer dollars to fund humans-in-space activities might well decline significantly, and space colonization efforts might well be undertaken privately. Currently, Elon Musk – founder of the space-access company *SpaceX* – publicly declares his ultimate goal to be to fund a colony on Mars, Burt Rutan and Richard Branson of *Virgin Galactic* have hinted that they are not interested in simply orbiting the Earth, but also want to support private expansion out from Earth, and Swedish–American explorers Tom and Tina Sjogren – who have walked to both of the Earth's poles unsupported and today administer the online *explorersweb.com* news outlet for exploration – are designing and raising support for their own, private voyage to Mars. As ever, there will be plenty who consider these people and their plans crazy and unrealistic, but just as surely the same charges were made at the Wright Brothers – and today aerospace industries are one of the main economic engines worldwide.

Ultimately, then, the idea that space colonization would cost too much money fails because it does not consider the following points:

- The essential human problems cannot be solved by money alone.
- It would not cost everything humanity has to give, and might well be cheaper than space flight and exploration to date.
- The costs might not have to come entirely, or even substantially, from taxpayers.
- The costs so far have been very small, and that there are other places to find ways to cut back on expenses, e.g. massive war expenditures.

---

[31] DeWaal (2009) succinctly reviews the overwhelmingly political drivers of modern famines.
[32] Zubrin (1997).

- A relatively high initial cost is often engaged in for a future benefit, as in the case of insurance.

We argue that humanity can afford to begin the process of space colonization at the same time that we address pressing issues on Earth. In fact, we argue, investment in space colonization would not be an expensive luxury, but a responsible investment.

Space Colonization Will Only Move Human Problems Off of Earth

In general conversations about the possibility of human space colonization, the idea that it would only result in human problems being exported to the next settlement is very common. In 2010, on the public debate website *debates.juggle.com*, the question was posed: "should humans attempt to colonize space?" Of the anonymous replies (respondents were 56% for and 44% against human space colonization) two responses characterize this negative position:

> "if earth is too crowded why should we just push our problem off to space and just over populate that? … we should find a way to stop using up our "renewable" resources quicker than they can replenish themselves. By people colonizing space its just trying to push our problem off on someone else."

> "It's bad enough we've ruined plant Earth and continue to destroy the planet and our value system because of our greed and selfishness. Of what benefit could it possibly be to anyone to colonize space or other planets? We're sure to continue on our course of mayhem and destruction. The human race isn't equipped to care for anyone or anything with integrity or intelligence."[33]

Essentially, these arguments suggest that humanity is incapable of learning or change, but this is clearly not the case: in the United States, for example, culture is significantly different in 2011 than it was in, say, 1951. In 50 years we have seen structural changes in the rights of women and minorities, built administrative structures to protect our water and food supplies from industrial wastes, and others to protect citizens from unsafe products. However imperfect these structures are (and of course there is much to improve) the point is that we have gone to the trouble to build them, indicating their worth to the citizenry. Much of this argument is personal and emotional, depending on the issue of whether one believes that humanity is largely good or largely bad. Of course, the truth lies somewhere between these two unrealistic opposites. Presuming that humanity is essentially bad is in our opinion fatalistic and myopically focuses on the bad that humanity has done, ignoring humanity's good acts.

---

[33] See http://debates.juggle.com/should-humans-attempt-to-colonize-space.

Having said all of this, it is of course true that humanity will carry its problems away from Earth. But if it is proposed that we solve human problems on Earth before migrating away, a significant problem arises: who is to declare, 'Now humanity is prepared for adaptation to space'? When will humanity recognize that fact, and what are its criteria? Perhaps this argument would be better served if such criteria were outlined; in fact, such delineation might provide a motivation to get on with real solutions to essential problems (outlined above) so that the migration to space could begin. We should also keep in mind that before solving the range of problems we might set out to solve, humanity could be struck with disaster, as described earlier in this book. And there is no guarantee that if some day humanity achieves the noble goal of global peace and prosperity, that it will persist; it could fail later.

For all of these reasons, it seems reasonable to work on immediate human problems while also thinking seriously about longer-term issues, such as space exploration. That can prove to be difficult, however, as pointed out by archaeologist Thomas H. McGovern and colleagues, who examined the failure of Viking colonies in southern Greenland in the 14th century AD. Although the cause of the Viking collapse remains specifically unclear, it is clear that it resulted from a mismatch of their subsistence practices to the Arctic environment, where they attempted to farm traditionally – as they had at home – despite increased cold and a declining weather pattern (authors' original emphasis):

> "... humans react not to the real world in real time, but to a *cognized environment* filtered through traditional expectations and a world view which may or may not value close tracking of local environmental indicators. Humans are also not always willing or able to forego short-term personal advantage for long-term common benefit."[34]

Our position is that of course we wish to improve humanity while on Earth, but that can be done at the same time that some of humanity begin to colonize space. Major points of human friction – religious intolerance, for example, or over-commodification of the natural world – are so deeply rooted that waiting to solve them before starting space colonization might, again, be penny-wise but pound foolish if plague, devastating war, or Earth impact by space object intervenes.

One of the more pervasive of human problems at large is warfare, an element of human life for many thousands of years.[35] It should be noted, however, that modern Earth warfare is extended over vast distances and long time periods and requires tremendous resources. At least early in the history of any extraterrestrial human colony, there will simply be too much to do – for example, in farming – and too few resources to make such conflict possible. In fact, conflict early in

---

[34]   McGovern et al. (1988): 266.
[35]   Many prehistorians and anthropologists have addressed the topic; good reviews are found in
         Guilaine and Zammit (2005), Kelly (2000), and Keely (1996).

human space colonies is much more likely to be characterized by the limited distance and limited duration of conflicts that characterize non-industrial cultures worldwide.

Still, it is often suggested by space colonization opponents that space will simply become the next battleground. That is of course possible, though so far the overt weaponization of space has been avoided by adherence to the 1967 Outer Space Treaty, whose central Article IV is summarized below:

> "This article restricts activities in two ways: *First*, it contains an undertaking not to place in orbit around the Earth, install on the Moon or any other celestial body, or otherwise station in outer space, nuclear or any other weapons of mass destruction. *Second*, it limits the use of the Moon and other celestial bodies exclusively to peaceful purposes and expressly prohibits their use for establishing military bases, installation, or fortifications; testing weapons of any kind; or conducting military maneuvers."[36]

Of course, treaties can be broken, but, again, one may always argue the worst-case scenario. And perhaps, if it is believed that humanity will never cure itself of war, that is even a stronger motivation to spread from just one planet, because whether anyone survives a full nuclear exchange on Earth, it can be questioned whether that life would be worth living.

Ultimately, the argument that humanity would only move human problems off of Earth is to a degree correct. However, that does not seem to be sufficient cause to abandon space colonization because:

- Waiting for humanity to perfect itself before colonizing space might take too long, if catastrophe strikes.
- It is unclear what humanity should achieve as a desirable state before it is mature enough to undertake space colonization (and, of course, it may be argued that even, once 'perfected', humanity could revert to prior behavior.
- It is possible for humanity to improve itself.
- It is possible for humanity to improve itself while at the same time making an incremental, responsible intellectual and capital investment in the colonization of space.

Space Colonization Would Simply Be An Unnatural, Technocratic Project

In *The Dark Side of the Moon*, Gerard DeGroot introduces the argument, penned by Lewis Mumford in 1969, that human space colonization would be an unnatural, anti-human endeavor, driven by motives far from those we have discussed in this book:

---

[36]   Outer Space Treaty (1967) at: http://www.armscontrol.org/documents/outerspace.

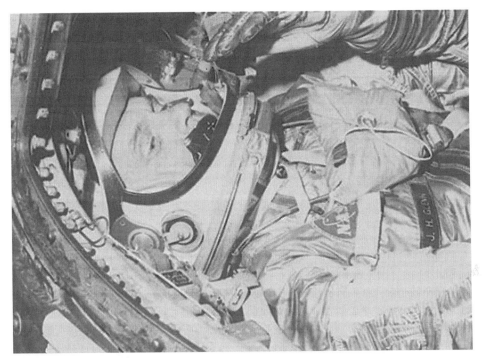

Figure 5.4. Astronaut John Glenn Being Sealed into his Spacecraft. Such imagery emphasizes the technology, making the human being almost an afterthought. NASA public domain image.

> "It is not the outermost reaches of space, but the innermost recesses of the human soul that now demand our most intense exploration and cultivation. Space exploration, realistically appraised, is only a sophisticated effort to escape from human realities, promoted by Pyramid Age minds, utilizing our advanced Nuclear Age technology, in order to fulfill their still adolescent – or more correctly – infantile fantasies of exercising absolute power over nature and mankind."[37]

This argument, once again, stands up a straw man: there is no reason that the innermost recesses of the soul cannot be examined while exploring or colonizing space, and it could easily be that new contexts and environments off of Earth would inform that worthwhile objective. The common charge of 'escapism', delivered here, can of course be leveled at anything that does not address immediate concerns; but remaining engaged only in the moment, we argue, is as irresponsible as escapism. Again, we argue that humanity can do several things at once, including improve itself and colonize and explore space.

---

[37] Lewis Mumford, who in the 1969 *Newsweek* article also referred to the Moon program as "an extravagant feat of technological exhibitionism", is quoted in DeGroot (2006): 267.

In the same vein, writing at the time of the Moon landings, Norman Mailer succinctly characterized one of the more alienating of the space race's many facets: technocracy. Commenting on the jargon-laden speech of the Moon-walkers, Mailer wrote that:

> "... everything in the astronauts' finicky man-supported, prone-to-malfunctioning, unhealthy, plastic, odor-sealed, and odor-filled environment was of equal news and interest ... [but the astronauts] existed ... like the real embodiments they were of technological man ... powerful, expert, philosophically naïve, jargon-ridden, and resolutely divorced from any language with grandeur to match the proportions of his endeavor."[38]

But, of course, Mailer's piling-on of adjectives is more an emotional reaction than factual; 'technological man' has no meaning in the study of humanity (anthropology), because technology is simply tool-use, which has characterized our genus for at least two and a half million years (as we saw in Chapter 2). Also, Mailer appears to have forgotten that on observing Earthrise from the Moon, Apollo astronauts read from *Genesis* while televising an image of Earth to the whole of planet Earth; and that on stepping on the Moon for the first time, Neil Armstrong clearly indicated that his single step was part of a much larger, inclusive, human endeavor. Still, Mailer was genuinely reacting to a real gulf between the astronauts and the general public, a wedge that astronauts themselves have commented upon. Michael Collins, Command module pilot for Apollo 11, has pointed out that astronaut communications were clipped and jargon-filled largely because most of the astronauts had been test pilots, people accustomed to delivering brief, dense, precise bits of information; and he has written that while the Apollo crews were technically the best people for the job, his 'dream crew' for space flight would be somewhat different, to overcome this exact hurdle:

> "The pity of it is that the view [of the Earth] from 100,000 miles has been the exclusive property of a handful of test pilots, rather than the world leaders who need this new perspective, or the poets who might communicate it to them."[39]

The argument, then, that space colonization would be a soulless, technocratic project, seems to be largely dependent on the idea that it would have the feel and philosophy of space *exploration* so far, which was, admittedly, alienating to some because of its heavy overtones of technocracy and even militarism, both of which were actively promoted by various space agencies to help justify their existence[40] (see Figure 5.4). But, as we argue throughout this book, space

---

[38] Mailer (1970): 212.
[39] Collins is quoted in White (1998): 37.
[40] McCurdy (1997): 218–221.

colonization is not space exploration; it is not an end, but a beginning, and it will not be about the few, but the many. It will not be like anything yet done in space. It will be about families and the proliferation of humanity, not about singular mostly-white, mostly-male, mostly-military people. The entire philosophy and feel, then, of space colonization cannot be technocratic; technology is only the tool to allow humanity to find new places for families to grow.

It is often suggested that space exploration's emphasis on technology would lead to a devaluation of space environments, such that Moon mining operations and the like would be irresponsibly carried out in the name of technical 'progress'. That might occur, but people have already begun to think seriously about how to be responsible in our exploration and colonization of space. In 1986, for instance, Sierra Club Books published *Beyond Spaceship Earth: Environmental Ethics and the Solar System*. Of course, one may always argue that humanity, served by powerful machines, will always decimate a given resource or environment. We feel that is simply taking the negative view, where a more mature recognition that much of humanity wishes to be responsible is more reasonable.

Ultimately, the argument that space colonization would simply be a technocratic and even militarized project fails because:

- Space colonization will not be the same thing as Cold-War space exploration, much (but not all) of which had different goals, philosophy and even language.
- Space colonization is ultimately about people, with technology serving only to make the adaptation to new habitats possible.

Space Colonization Would Be Immoral

At least two dimensions of morality regarding human space colonization have so far been seriously discussed. One is the issue of consent to risk, in which one may ask whether space colonists would ever be placed in environments where their ability to reject risk was constrained. If so, one might say that the planners who put those colonists into that position had acted immorally. In an examination of this issue, American philosopher Norman Daniels has pointed out that over time, colonists, historically (and presumably in the future) have themselves changed over time, leading to new concepts of risk, and that, ultimately, consent to risk would actually shift as an issue from Earth-based administrators to off-Earth colonists themselves. In other words, while early on in space colonization it would of course be important to make hazards known and provide the freedom to consent to undertake or reject those risks, after some time the moral issue would dissolve as colonists diverge culturally from Earth.[41]

The second, related, moral issue addresses the issue of whether children of space colonists would be treated immorally in that they did not consent to being

---

[41]   Daniels (1986).

born into colonial conditions. For example, in a paper on issues involved in population growth in extraterrestrial habitats, William A. Hodges has painted a gloomy picture of pioneer life for space colonists, writing that "... for generations the settlers will be poor. They will spend their entire lives in conditions of extreme material deprivation and high risk of death," continuing that if such colonists lost individual non-food production specialists, life would become dim, quickly; if the colony's cobbler died, Hodges suggests, the people would have to go barefoot.[42] This strange perspective ignores the simple remedy of passing the cobbler's skills on from one generation to the next, as has been done for thousands of years on Earth! More seriously, this perspective ignores the essence of human culture itself; that it is cumulative and transmitted from one generation to the next by social transmission. More importantly, the argument that it would be immoral to bring children into a world involving risk is unrealistic; risk attends any life. Regarding the material conditions into which children are born, it is immediately evident that the entire set of material conditions of modern, Western civilization are not necessary for a happy existence; many human cultures over thousands of years and across the globe have had rich and fulfilling lives without, for example, three cars per family. So, exactly what constitutes 'deprivation' in such an argument must be clearly defined. It is tempting to argue that millions of humans today are born into conditions of terrible poverty and an uncertain future, and that for this reason space colonies' doing the same would be justified, but this assumes that the act is moral simply because it is widespread, an untenable position. Certainly, moral parents will always want to provide a healthy and relatively safe environment for their offspring, but we should consider that in the developed world today we might be laboring under the illusion that if conditions for life in a given space colony are not identical to our own, they would be 'bad'. It could also be argued that, given the opportunities of life off of Earth, humanity would be acting immorally if it did not undertake space colonization as an insurance policy for future generations.

Some might argue that being born outside of Earth culture at large, humans might be deprived of a religious tradition and, therefore, a code of moral behavior, again a condition rooted in the immoral act (in this argument) of setting certain conditions of the life and environment of space colony children. But this argument takes the untenable position that only religious tradition can devise guides to moral behavior, which is clearly untrue when one considers the results of the field of moral philosophy. For example, philosopher Bernard Gert has proposed the secular concept of common morality by which human cultures strive to avoid five harms: death, pain, disability, loss of freedom and loss of pleasure. This requires attention, Gert argues, to ten secular rules:[43]

---

[42] Hodges (1985): 140.
[43] See Gert (2004).

- Do not kill.
- Do not cause pain.
- Do not disable.
- Do not deprive of freedom.
- Do not deprive of pleasure.
- Do not deceive.
- Keep your promises.
- Do not cheat.
- Obey the law.
- Do your duty.

Ultimately, arguments that suggest that human space colonization would place human beings into risk environments, and constrain their freedoms, are weak because:

- They fail to clearly identify what constitutes undesirable material conditions of life.
- Colonial cultures will eventually diverge from Earth cultures in their evaluation of moral issues.

When we consider the first colonists to leave Earth, it is easy to become overwhelmed with the risks and deprivations to which it seems they might be exposed. But we can be sure that all will be volunteers, and we should recall that our own migration experience in the developed world has only recently faded, as Arthur C. Clarke pointed out in 1960:

> "Only a lifetime ago, parents waved farewell to their emigrating children in the virtual certainty that they would never meet again."[44]

Conclusions

It is easy to dismiss the concept of space colonization by invoking technical, biological, ideological and moral issues. It is equally easy to advocate for space colonization by simply papering such issues over. A more measured and mature approach, as we have shown in this chapter by dissecting such arguments, indicates that many of these arguments are not well prepared. None, at present, clearly bar human colonization of space technically, biologically, culturally or morally. In clarifying these issues the actual motivations and philosophy of a mature, considered attempt at human space colonization have become clear. It is not about the conquest of space, or the advance of technology over humanity, or planting a flag on the Moon, and so on. It is about opening a new niche for humanity – and all of our coevolving species – to continue to evolve, a gesture that is in harmony with the nature of the Universe itself: not fixity, but change.

---

[44]   Clarke (1999): 205.

Part III

Human Adaptation To Space: Lessons From The Past And Plans For The Future

6 Starpaths: Adaptation to Oceania

"Not only do [the Polynesians] *note by* [the stars] *the bearings on which the several islands with which they are in touch lie, but also the harbours in them, so that they make straight for the entrance by following the rhumb of the particular star that rises or sets over it: and they hit it off with as much precision as the most expert navigator of civilized nations could achieve."*
18th century Spanish mariner[1]

Equipped with a general understanding, now, of how humanity has decoupled behavior from anatomy by using tools – and thus changed the course of its own evolution by *inventing invention* – we can go on to see two specific cases of proactive human adaptation and niche construction in action. There are plenty of examples we could use, but here we will focus on the example of the Polynesian colonization of Oceania, the Pacific at large, from Australia to the Hawaiian Islands, New Zealand, and Easter Island. This case provides superb examples of invention and adaptation that can inspire, provide material for creative thought, provide new vocabularies to help reconsider human space colonization, and, overall, energize the project of human space colonization. We will see the construction of great sailing vessels, often as large and complex as those of the European ships of discovery, starpath navigation and an ethos of exploration and discovery, all devised not as ends in themselves, but as adaptations to make human colonization efforts a success.

Not only will we find technical prowess, but we see cultures built around the principle of exploration and the adaptive flexibility required by the expansion into new ranges. While thinking about and planning human space colonization, we can find inspiration in the Polynesian voyagers who migrated into vast, unknown and forbidding environments, equipped with tools that today, naively, we might call crude. We will see that all are simple technologies in comparison with spacecraft, but each is a masterpiece of adaptive design. And we will see that just as important to possessing the technology to make this colonization was the fact that Polynesian cultures highly valued exploration for its own sake – something Western civilization at large might have difficulty saying of itself today.

---

[1]   Corney (1919).

Oceania: Initial Colonization of the Western Pacific

As we saw in chapter 3, behaviorally-modern humanity emerged from Africa on the order of 100,000 years ago and subsequently swiftly spread into many regions; Europe, Siberia, the river plains and jungles of sub-Himalayan Asia, Australia, the Americas. In each area, distinctive cultures evolved, their tools, customs and languages shaped – at least in part – by the ecological conditions. The settlement of Austronesia (Australia and the archipelago extending from it to mainland Southeast Asia) required watercraft to traverse offshore waters in stretches of up to 50km, beyond the sight of land in any direction. We have no archaeological evidence what these watercraft were built of or looked like, but a long tradition of lashed-bamboo watercraft in the region provides a possible hint. Whatever their form, such vessels allowed humanity to colonize Australia, and, pushing further east and north, the islands of New Guinea, Borneo and Sumatra, all certainly by 50,000 years ago. Figure 6.1 shows this vast region.

Not long after this colonization, some humans also pushed eastwards into the practically countless islands of Melanesia, on the far southwestern margin of the Pacific. By the time European explorers arrived here in the 1500s AD, large native sailing vessels were common, but again we do not know how they relate to the earliest craft of the region. What we do know is that the watercraft were common and capable of inter-island navigation.

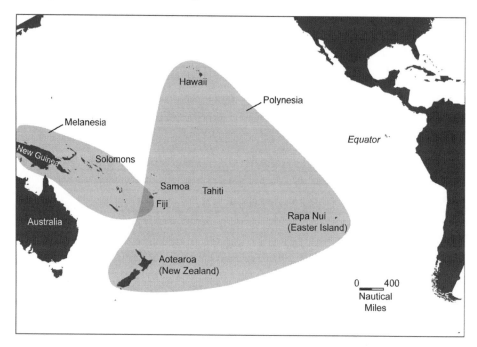

Figure 6.1. Polynesia and Melanesia, Collectively Referred to as Oceania. Map by Cameron M. Smith.

Figure 6.2. Dr Evan Davies Excavating a Lapita site on New Ireland. Photo from the collection of Evan T. Davies.

The Lapita Culture

Despite these advances, successful, long-term navigation on the open ocean, pushing yet further east into the Southern Pacific, was not attempted for many thousands of years. When it did, it was first carried out by an archaeological culture referred to as the 'Lapita'.[2] In 1988, one of us (Evan T. Davies) worked directly with archaeologist J. Peter White of the University of Sydney, Australia on an excavation on the island of New Ireland off the coast of mainland Papua New Guinea. Professor White spent his career uncovering many details of the prehistory of Melanesia and New Guinea, and shed a great deal of light on early migrations throughout much of the Western Pacific. On New Ireland (see excavation in Figure 6.2) that winter we investigated aspects of the Lapita culture

---

[2]   A good overview of Lapita culture archaeology is found in Kirch (1997). Our review of Lapita and post-Lapita colonization of the Pacific uses archaeological (see Golson 1977 and many other references in this chapter, including Webb 2006), DNA (see for example Merriwether et al. 1999 and Kayser et al. 2000), lingustic, historical and oral tradition, some of which is synthesized in Finney (1994). It is interesting to note that Polynesian oral traditions are divided by Buck (1938) into a mythical period regarding the origins of the Polynesian world, and a history of the colonization of the islands of the Pacific.

and its antecedents, and gained first-hand insights into the remarkable culture of these early Pacific colonists.

Originating roughly 3500 years ago, the Lapita culture is recognized by its distinctive artistic styles as well as a suite of artifacts, including distinctively shaped fish-hooks, a sophisticated pottery technology and skilled drilling, perforation and grinding techniques to work a wide range of materials, including wood, shell and bone. Most strikingly, traces of Lapita material culture are quickly found outside their Western Pacific area of origin, reaching as far east as Tonga and Samoa by 3000 years ago. This wide distribution, accomplished in just a few centuries, argues strongly for the Lapita having been proactive colonists. While some early settlement of the islands of Melanesia might have been accidental, as westerly winds in that area might naturally blow sailing craft downwind (east) onto islands, Lapita culture expanded rapidly *into* the prevailing easterly to the east of Melanesia. Although wind directions at sea can change through time, it has been shown that in the last three millennia the winds would have been essentially as they are today: largely easterly, blowing against anyone wishing to move east into the South Pacific.

Archaeological excavations throughout the South Pacific have given us a good sketch of Lapita culture. We will turn to their ships and navigation methods a bit later in this chapter, but for the moment we can learn something of their basic way of life.

Lapita Subsistence

In terms of subsistence, Lapita people combined the foraging, hunting and collecting of food resources that had been used by hunter-gatherers for millions of years beforehand, but they were also domesticators. Domestication, we saw before, is the capture of non-human species for human use; it includes a spectrum of control and intensity from *horticulture,* carried out on a rather small scale and storing its products for only months to seasons at a time, to *agriculture*, which uses massive irrigation works, intensive crop rotation, and stores its products for years at a time in large facilities such as granaries. Lapita people were on the horticultural end of the spectrum, farming such plant foods as taro and yams, and animals such as pigs and chickens, all species derived from Southeast Asia.

Additionally, of course, they hunted and collected every imaginable resource from the ocean, including fish of all kinds (from minnow-sized to tuna-sized), shellfish (including lobsters), mollusks such as clams, and birds. Specialized equipment, including nets and all manner of fish hooks, gigs and so on, were invented, each a material adaptation to the oceanic environment.

Lapita Raw Materials and Tools

Many plant and tree species were also carefully managed not for food, but to provide raw material; their ships were no simple dugouts, but elegant craft built with as wide a variety of materials as any European galleon. In the same way, building materials were collected from animals, including mollusk shells hafted into handles as fine woodworking adzes, shark skin used as 'heavy grit' sandpaper, small conical snail shells used as perforation drill bits, and animal bones (sometimes the bones of their own ancestors) sharpened in the manner of the finest wood chisel. Commenting on Lapita descendants he encountered in the Pacific in 1769, English explorer Joseph W. Banks noted:

> "Their tools are made of the bones of men, generally the thin bone of the upper arm; these they grind very sharp and fix to a handle of wood, making the instrument serve the purpose of a gouge by striking it with a mallet made of a hard black wood ... they hollow with their stone axes as fast at least as our Carpenters could do ... I have seen them take the skin off of an angular plank without missing a stroke, the skin itself scarce 1/16th part of a inch in thickness ... With these boats they venture themselves out of sight of land; we saw several of them at Otahite [Tahiti] which had come from Ulheitea and Tupia [a Polynesian navigator] told us that they go voyages of twenty days ... they keep [the ships] under boathouses ... one of which we measured today 60 yards by 11."[3]

Today, some few native Polynesian islanders on the island of Raivavae (about 500km southwest of Tonga) continue to use essentially Lapita materials and methods in building their ships. One late 1990s anthropological survey noted the wood and other plant materials used in the construction of just one vessel, summarized in Table 6.1 (which is adapted from Table 2, p. 109 of Vecella (2008)).[4]

In the same way, Indonesian traditional shipbuilders, today known as Bugis, were recently observed to use the coconut palm for multiple functions:[6]

- Young root = glue for binding stone to wooden anchor shaft.
- Trunk wood = houses, furniture, ship's bulwarks.
- Bark = paint brushes.
- Leaf midribs = walls and fences.
- Whole or split leaves = thatched roofs, walls, dugout covers.
- Leaves = woven into mats to line containers, dry fish and panel house walls.
- Immature fruit = snack.
- Coconut: mature nut/grated = wood treatment, food, cooking oil.
- Coconut: grated flesh after milk is extracted = chicken feed.
- Coconut shell = water vessel, charcoal.
- Coconut husk = cordage; brooms; furniture stuffing; cleaning scrubber; toy boats.

---

[3]　Beaglehole (1955): 154.
[4]　See Vecella (2008).
[5]
[6]　See See Ammarell (1999), Table 2.1, p. 27.

Table 6.1. Natural Materials Used in Construction of Traditional Polynesian Sailing Vessels.

Common name	Polynesian name	Use
Beach hibiscus	Purau	Wood for all parts of vessel, esp. nose (bow cover), outrigger float, forward crossbeam, paddle, bailer, and mast. Bark is converted to cordage and rope.
Coconut	Tumu ha'ari	Husk fiber for lashings. Oil for caulking. Nutshell to block lashing holes. Stem and leaf fibers for sails.
Alexandrian Laurel	Tamanu	Bailers and hull
Fish-poison tree	Hoto or hutu	Wood for bow and stern covers, fruit fibers for caulking and hulls
Mango	Tumu Vi	Wood for bow and stern covers
Breadfruit	'Uru	Sap for caulking and glue
Ironwood	'Aito	Rear crossbeam
Pacific Rosewood	Mimro ou amae	Dowels and bailer
Jacquin	Apiri	Rear crossbeam
Pandanus	Fara	Plaited mat sails
Unknown	O'iri	Seed juice decorates lashings
Banyan	Aoa or ora	Cordage
Unknown	Hoto	Paddles

One important lesson here is that Lapita raw materials were often multi-functional; another is that they were renewable. Lapita people, in fact, carried seed stocks on their voyages to plant vegetation not just for food, but also for building materials. A second lesson is that, of course, like all pre-industrial people, the Lapita were conservative and re-used materials in a *curated* ethos in relation to their resources rather than a wasteful, *expedient* relationship to those resources. At least initially, space colonists will not be able to be wasteful; the price for materials will simply be too high. In the same way, scattered across islands in many millions of square miles of the Pacific, Lapita people commonly reused their artifacts when 'broken', such that 'broken' pottery was not thrown overboard, but repurposed as a bailing tool, or a protective container for glowing cinders. It is heartening that a large-scale, paradigm-level adoption of such 'green' use of resources (as mentioned in Chapter 7) is already underway in the fields of architecture and civil engineering.

In addition to naturally-occurring materials, the Lapita transformed certain resources by heat and other methods into new materials. Clay was mixed with crushed shell and a variety of sands (each shell/sand mixture specific to a certain pottery function, such as storage of water or food), shaped by hand and then heated in open bonfires to create pottery. Pottery was critical to storing and transporting food, water and other supplies, as we will see when we come to Lapita voyaging ships.

The Lapita people, like all humans, also extensively used stone: coarse volcanic ash (pumice) was used to abrade wood; finer-grained lavas (e.g. basalt)

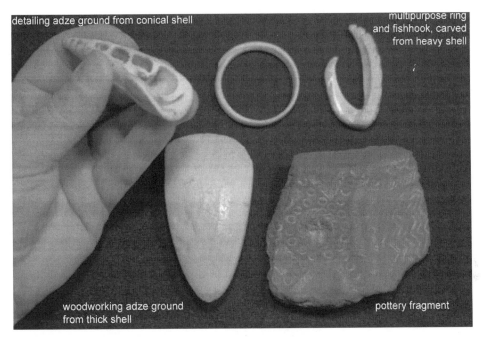

detailing adze ground from conical shell

multipurpose ring and fishhook, carved from heavy shell

woodworking adze ground from thick shell

pottery fragment

Figure 6.3. Distinctive Lapita Artifacts. Photograph from the collection of Evan T. Davies.

were chipped into heavy woodworking tools or abraded into fine woodworking axe heads; cherts (e.g. jasper or flint) were used for cutting tasks and could be finely shaped into precision tools; and volcanic glass (obsidian), blades of which are used for surgery in modern hospitals, was so valued that one study shows it was transported over 2500km from its source volcano.[7]

As mentioned, shell was worked into bracelets, pendants, fish hooks, chisels and 'coconut graters', which preprocessed coconut meat; Figure 6.3 shows some common Lapita artifacts.

The Lapita used, of course, a wide range of other materials, including mats of plaited fibres as sails on their vessels and roofs for their houses, but only scraps of these remain today.

Lapita Social Organization

Archaeologists have a long history of reconstructing ancient cultures' social organizations by 'reading' the artifacts. Socially ranked cultures, for example, are relatively easy to identify because they often use artifacts to conspicuously display wealth, as in the case of a Polynesian chief's elaborate head-dress

---

[7]   Ambrose and Green (1972).

(compared to a commoner's hat), or a modern person's choice between driving a BMW or a Volkswagen (both cars will get you from A to B, but each transmits something about its owner to other people in the culture).

Lapita artifacts show that the culture was somewhat socially ranked, and its character seems to have significantly shaped the Polynesian peoples, often arranged in powerful, competing chiefdoms, that European explorers encountered when they first came to the South Pacific. Generally speaking, Polynesian chiefdoms, and by extension, their Lapita ancestors, were arranged in populations of up to around 2000 people under the general sway of a chief. Chiefs were royalty, born of prior chiefs or other royal blood, and their families, like the royal families of Europe for instance, could maintain decades or centuries of power before being replaced by members of other lineages (the end of one *dynasty* and the beginning of another). The archaeological traces of these powerful chiefs (by the time of the European arrival, some male, and some female) are their larger-than-usual houses and their specially-carved bone wealth-display artifacts, such as ceremonial clubs as precious to their cultures as the sword 'Excalibur' was to ancient English people). Through time, some chiefs were certainly more self-aggrandizing and self-interested than others, and others were more magnanimous, acting more as leaders than rulers. Whatever the case at any given moment of Lapita history, we'll see later that Lapita social organization might have been instrumental as a motivator for colonization of the Pacific.

Like many ancient societies, Lapita also conveyed their social organization in ways that have not survived in the archaeological record, but can be inferred from practices of their Polynesian ancestors. Many Lapita artistic design elements visible on their pottery, for instance – including repeated geometrical patterns, narrow parallel lines in broad series, stylization rather than naturalistic depiction or portraiture – are seen in the distinctive tattooing of Polynesian royalty (see Figure 6.4) known to early Pacific European explorers.

Essentially, Lapita peoples appear to have been maritime-adapted horticulturalists arranged in royalty-managed village populations. Cities and states of the size and complexity we know from the ancient civilizations, such as the Inka or Egypt, did not develop. This is very likely because such large entities as civilizations and the massive populations they support (millions, for example, in ancient Egypt), require intensive agriculture and storage facilities, neither of which were possible on the relatively small Pacific islands. We must remember this 'limitation' (though it is only a limitation in hindsight) when reading early European descriptions of Polynesian culture; Polynesian chiefs were no less driven to build their political sway than any Egyptian pharaoh, but worked within the boundaries of their environmental conditions. In the same way, the shape of space colonies will not 'naturally' evolve from one form to another, based on some internal engine that drives human social organization through predetermined stages; such stages, anthropology and archaeology have shown, do not exist. Human cultures *have* in the past changed in rather predictable ways (again, in hindsight), but they do not do so universally or according to any internal engine.

Post-Lapita Voyagers and Colonists of the Pacific

Lapita culture made the first great expansions east from Melanesia, colonizing, as mentioned, as far as Tonga and Samoa by 3000 years ago. Figure 6.3 shows some distinctive Lapita artifacts. After this time, however, the distinctive Lapita archaeological culture wanes. It appears that the final wave of migration throughout the rest of the Pacific occurred after about 1500 years ago, by direct ancestors of the modern Polynesians. This colonization wave was particularly rapid and purposeful; again, both practical observation of the prevailing winds in the area, and computer simulations (one running 100,000 simulated 'accidental drift voyages') of what would happen to vessels that became lost or disabled show that they could not conceivably have simply drifted, hap-

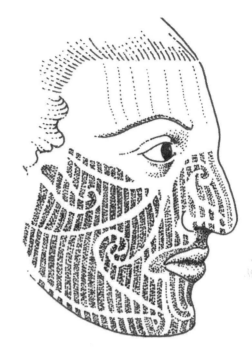

Figure 6.4. Typical Polynesian Tattooing. Drawing by Cameron M. Smith, after a 19th century engraving.

lessly, to the islands of the Pacific, but must have sailed against the wind to reach them.[8] The rest of this chapter describes how that was done, and on what kinds of ships.

Traditional Polynesian Sailing Vessels

Much of what we know about the ships used in this most proactive colonization effort comes not from archaeology, but from early European explorers'

---

[8] See Levison, Ward and Webb (1973) and Webb (2006): 113. In other research, Irwin has modelled the sailing capacities of traditional Polynesian vesses and set them in his digital ocean, replicating radiocarbon-date-indicated process of colonization of the Pacific islands (See Irwin 1992, Irwin 1998, Irwin and Blacker 1990), using the known Polynesian method of maximizing potential return downwind if nothing was found upwind, a tactic noted by Lapita researcher Ben Finney. Another simulation using bootstrap methods indicates that a voyage from Pitcairn to Rapa Nui (Easter Island) would have averaged 17 days, a figure corroborating the 1999 voyage of the experimental Polynesian ship replica *Hokule'a* from Pitcairn to Rapa Nui, which took 14 days.

descriptions of what they encountered when the arrived in the Pacific. As early as 1579 Sir Francis Drake commented that Polynesian sailing vessels hulls were 'very smooth within and without',[9] but more detailed descriptions came with 18th-century explorers such as Banks and Captain Cook. To read these accounts, while some voyagers were favorably impressed by the Polynesians, we often need to remember that ethnocentrism and European plans to annex the Pacific islands for their own territories often combined in ways that denigrated the Polynesian cultures. One French explorer, Dumont D'Urville wrote of the Western Polynesians:

> "These blacks are almost always grouped in very weak tribes ... Far more debased towards the state of Barbarism than the [Eastern] Polynesians and the Micronesians, they have amongst them neither a regular form of government, nor laws, nor established religions and their intelligence are also generally much inferior to those of the copper-colored race [the Eastern Polynesians]."[10]

Of course, no society is without laws or established religions, and our review of the cognitive foundations of behavioral modernity demonstrated that complex symbolism and language – both possessed by all humans – are at the actual root of 'intelligence', which cannot be adequately described in such pronouncements.

With regard to the Polynesian ships themselves, we must overcome a similar significant hurdle resulting in this kind of historically-contextualized description. Early on, Spanish explorers used the term 'canoa' (=canoe) to describe Polynesian sailing vessels, seemingly grudgingly admitting from time to time that they were often as long as Spanish vessels themselves, carried plenty of sail and were, as we will see, as capable of blue-water (offshore) voyaging as European vessels. As an example, consider Table 6.2, a rough guide to Polynesian (and some other) ancient oceangoing vessels:

Table 6.2. Some Properties of Sailing Vessels of the Ancient World.

Origin	Ship	Years Ago	Length (feet)	Tonnage	Passengers & Crew
Polynesian	Double 'canoe'	>2000	80–110	c.40–60	50–60
Viking	Gokstad	900	76	20	c.70
Columbus	Santa Maria	500	80	100	39
British	Mayflower	350	90	180	150
Capt. Cook	Endeavour	200	106	368	c.85
Ecuadorean	Sailing raft	1200	60	20	30

---

[9] Alexander (1916): 114.
[10] d'Urville (1832): 11.

Indeed, the first concern of many authors summarizing what we know of Polynesian voyaging ships is to point out that the terms 'canoe', or even 'double canoe' or 'raft' – anything other than 'ship', actually – is simply an ancient and inappropriate result of archaic, ethnocentric terminology. As late as 1939 a 100-foot traditionally-built *baurua* (voyaging ship) was photographed in the Gilbert Islands, a vessel clearly built not for close inshore fishing (!) but for long-distance, 'blue-water' voyaging.[11] Polynesian voyaging vessels were ships. We should also keep in mind that while Polynesian exploration ships could reach over 100 feet in length, such large vessels (even in the European tradition) are almost always too bulky for the exploration of islands; even the Spanish conquistadores regularly used smaller caravels, of 60–70 feet (and massing about the same as Polynesian ships) for their exploration of the New World, in contrast to the larger, more ponderous galleons used once trade routes had been established.[12] This indicates that overall dimension is a poor measure of a ship's utility for exploration.

Another term to be aware of is 'outrigger', which evokes the idea of a small boat like a catamaran, or a canoe with a pole extending to one side, with a float on the far end. Certainly some Polynesian vessels had exactly this form, but the large voyaging vessels were not simple outriggers, they were ships. Throughout this discussion, Figure 6.5 will be a useful reference.

Traditional Polynesian ships were built for all kinds of functions, just as we distinguish between tugs, cargo ships, lighters and so on. All were based on two main elements: sails and a hull. Again, historical accounts tell us much of what we know of the ancient design.

Sails were woven from vegetable materials, and were often replaced yearly. Explorer William Ellis wrote of Polynesian sails he observed in 1829:

> "... The sails were made with the leaves of the pandanus split into thin strips, neatly woven into a kind of matting. The shape of the sails of the island-canoes is singular, the side attached to the mast is straight, the outer part resembling the section of an oval ... "[13]

Like European sails, Polynesian sails were affixed to various yards and masts, and were controlled with lines made from twisted or braided cordage (European cordage was often made of hemp or other fiber, imported sometimes from half a world away). One distinction of some Polynesian vessels was that their masts were not fixed in one place, in the European manner, but were mobile, such that they could be quickly and easily moved from one end of the vessel to another. In fact, one common way to change the ship from one tack to another was to simply turn the ship into the wind, move the mast and sail to the required position, and resume sailing on the opposite tack; and some vessels' tracks could

---

[11]   See Plate IX, p.262 of Lewis (1972).
[12]   Smith et al. (2006).
[13]   See William Ellis quoted in Dodd (1972): 137.

Figure 6.5. Typical Polynesian (top) and European Voyaging Ships. Image by Cameron M. Smith.

be quickly *reversed* by moving the mast from the 'front' to the 'rear' of the ship, such that there was not necessarily a formal idea or design that favored one end of the vessel or another.

The hulls of Polynesian vessels differed tremendously from those of European vessels. Larger Polynesian vessels had two hulls, not one; these were formed by building planking up around a basic dugout log, tapered at either end and shaped below waterline into a hydrodynamic taper. Giant dugout hulls, up to 30 meters (100 feet) long, or even longer, were based on trees felled with stone axes (numerous experiments have shown that stone axes are as effective as steel axes at felling trees, though they are more difficult to sharpen and can break more easily). The selection of trees was a gravid matter, as in any ship construction, and the building process could take years from tree selection to finished ship. The tree-felling area was carefully selected,[14] such that large hangars could be built to protect the ship as it was built on-site.

Once the hull trees were hollowed out, planking was attached, often by sewing cordage through strategically-placed perforations in the planks, and the use of simple tourniquet devices to adequately tighten the joint. If these building materials and methods sound primitive, laborious, or unusual, one might recall that the space shuttle's tiles are glued to the hull, one by one, and that the thousands of individual tiles are so delicate that they can be scratched with a fingernail. Planking was built up around the dugout log cavity, and sealed at the top, to prevent water from entering the compartments and to make room to store provisions, such as food and water containers.

The hulls were joined by crossbeams of particularly durable woods, and a living platform, often rectangular, was built upon these crossbeams. Often, a one-or two-storey wooden structure, just like a galleon's deckhouse, was built on this deck, to house the galley and other vulnerable facilities. The galley contained stone slabs and sandboxes in which to build and maintain fire for cooking and other purposes. The roof of the deckhouse was often used by navigators as an observation platform. With large crews, entire families were on board, and would sleep in compartments in the covered hull dugouts. Such compartments could not have been less comfortable than those on European vessels. It has been pointed out that the wide stance of the Polynesian double-hulled ship makes it much more stable (resistant to seasickness-inducing motions) than a typical round-bottomed European vessel, and that this stability was an advantage in starpath navigation, a method we shall review below. Whatever the case, it also, certainly, made life more pleasant for any passenger.

Final building stages included caulking the planking (sealing up gaps between planks) in a way functionally identical to that of European vessels, using various sticky substances, often kneaded into fibrous material, hammered into gaps between planks. Cordage was a big concern, as on any ship, and even a 30-foot

---

[14]   Vecella (2008): 112; see also Sinoto and McCoy (1975).

vessel, today, requires at least 100m (>300 feet) of quality rope. One study of traditional cordage production on the island of Truk found that 10 old men, and 47 younger men, produced 4065 feet of good, construction-grade cordage made from the fibers of a nettle-like plant twisted into rope for the building of a large house.[15]

The importance of shipbuilding was paramount to the Polynesian explorers, and it is encoded in a traditional Hawaiian veneration of the adze, the stone tool used to hew the hull:[16]

> Go and dig out the adze
> In the adze pit in Havaii
> Hold, that it be taken out and enchanted
> Made light; that it may shoot sparks
> In doing its work
> It is whetted with fine sand
> Made smooth with loose-grained sand
> It is set in a firm handle of sacred miro
> Bound with many-stranded sennit of Tane
> The adze will become sacred
> In the brilliant sennit of the artisan
> Which touches and holds
> As a girdle for the adze
> For the handle of the adze
> To the back of the adze
> To make one the adze and the handle
> To make light the adze
> To consecrate the adze
> To impel the adze
> To complete the adze
> To give power to the adze

The building of a ship might be carried out by up to ten workers overseen by an experienced shipwright, and finished vessels could well exceed 40 tons in weight. As mentioned, once at sea, sails and other elements would break down over time, but this is not a problem unique to Polynesia; anyone who has crewed any wooden craft knows that repairs and adjustments are a constant of life aboard. They are not a problem or a disaster, and spare parts, and the knowledge to 'jury rig' solutions, are always carried in large quantities.

Polynesian voyaging ships, then, were large, sturdy, massed in the tens of tons, could carry (easily) up to 50 people at a time (say, 10 four-person families of a married couple and two children each, plus 10 others), could easily carry many

---

[15] Buck (1950): 51, 123–141.
[16] Dodd (1972): 107.

tons of supplies, and were as durable as any European craft.[17] They were ships of colonization.

Polynesian ships could also carry out all basic maritime maneuvers at sea: turning, tacking (sailing against the wind as capably as European vessels), anchoring, getting offshore, docking and so on. Captain Cook noted that a standard double-hulled *Pahi* (voyaging ship) could outpace his own *Endeavour*, to voyage, easily, 100–120 miles daily.

Traditional Polynesian Navigation

In the same way that the nature of Polynesian vessels were inappropriately referred to as 'canoes' or other insubstantial vessels, their navigation techniques must be reconsidered. Again, E. Dodd reminds us, in his masterful treatment of the topic, that:

> "We must cease trying to figure out how close the Polynesian came to our [European] navigational system, especially how he might approximate latitude and guess at longitude. We must seek out an entirely different system, a basic, comprehensive one that is conceived in his terms as a Polynesian, not in ours as a European ... the Polynesian had no angles or degrees; he did not think that way. What we must do is throw overboard the whole notion of a latitude and longitude grid. It is patently non-Polynesian. The Polynesian never thinks in terms of lines or abstract intervals of measure. His [units include] the breadth of the hand, the digital span, the divided forearm, the finger tip to elbow ... Distances ... he measures by the time it takes to negotiate them ... Especially he does not think in terms of arbitrarily placed and spaced lines that must be expressed in graphic markings on flat paper ... He does not place his islands by their relative positions north and south, east and west. He places them in the direction where they lie from where he is standing. [renowned European navigator Harold] Gatty says [the Polynesians had] 'no word for distance, no use of space.'"[18]

In short, the European system of navigation is based on knowing one's own position on the Earth in terms of the extremely useful latitude and longitude grid, based on the equator and the prime meridian at Greenwich, England, respectively. For a long time, this position has been identified by the use of a sextant to identify latitude and a chronometer to help identify longitude. In contrast, the Polynesian system of navigation *theoretically keeps the vessel stationary, and considers the position of other things around it.* These other things,

---

[17]   Lewis (1972): 273–274.
[18]   Dodd (1972): 42–43.

such as the various islands, are thought of (not literally) as sliding past the vessel. This method, as we will see, makes it much easier to use the stars and other markers to identify one's position. Many traditional Polynesian navigation methods were almost lost by the 1960s, but were rescued when anthropologist/ mariner David Lewis traveled to Polynesia and sailed thousands of miles with traditional navigators to learn and record their methods; more recently, the Hawaii-based Polynesian Voyaging Society has revitalized the ancient methods, and teaches them formally to both interested Europeans and native Polynesians alike.

Two main tasks were carried out in Polynesian navigation and seafaring: knowing one's position, and finding land.

Identifying position was often done with celestial cues, as visually explained in Figure 6.6. In this image, the sky is calculated for its appearance at 10°N 140°E (in the Caroline Islands) on Sat, 23 April, 2005, at 11:00:00 UTC (02:00:00 Local Time). Only visible Carolinian navigation stars are shown. Star 1 is Kochab, with a rising azimuth of 20°: in the figure it is at 16° above the horizon and at azimuth 15°. In the same way, star 2 is Dhube, which rises at azimuth 30° and is now at 37° above the horizon and at azimuth of only 6° from the bow of the vessel. Star 3 is Aldebaran, which rose at azimuth 70°, is now 5° above the horizon, and at azimuth 286°. Alnilam, in Orion's belt, rose at azimuth 110° and is now at 17° above the horizon and at azimuth 266°; and Acrux, having risen at azimuth 150° is now 13° above the horizon and at azimuth 167°. In Polynesian voyaging, of course, azimuths are not measured with degree-reading instruments, but are estimated with, for example, the fingers. In this image, A is the current course and B the desired course of, say, 25° east of Polaris, the northern pole star. With the present compass stars visible, the vessel would be steered between Kochab (1) and Dhube (2), which rise at 20° and 30°, respectively. Kochab, around 16° above the horizon, is just becoming useful as a directional star. Dhube, at almost 40° above the horizon, is less useful, until it nears its setting azimuth of roughly 310°, because rather than keeping it in sight on the bow, when it is high in the sky one has to look up at it, and then back down to where the vessel is pointed. For the moment, the vessel could be trimmed towards Kochab, but a bit south of it. Acrux (5), now close to 15° above the horizon, would be kept on the starboard quarter, with setting Aldebaran (3) and Alnilam (4) a few degrees aft of the port beam. As directional stars set, others rise. The navigator would monitor the relative position of the stars (without instruments) to maintain course.

In short, the rising and setting of the stars were used to find one's way. At the equator, for example, the constellation that we call Orion always arises at a certain time (depending on the season) directly to the east and sets – hours later – directly to the west, just like the Sun. If one sees Orion rising to one's starboard quarter (the rear, right corner of the vessel), one knows that one is travelling roughly north. If it rises from the port quarter (the rear, left corner of the vessel), then one knows one is travelling roughly south. If one is traveling north, and wishes to travel somewhat east, one may turn the vessel somewhat to the right, keeping Orion somewhere off the right (starboard) side of the vessel; and the

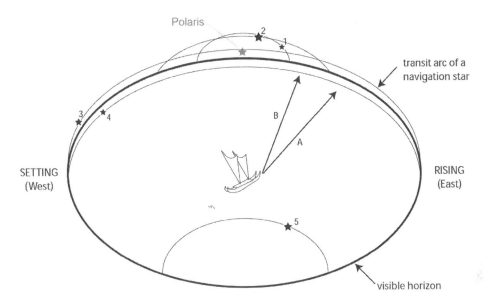

Figure 6.6. Starpath Navigation. Image by Cameron M. Smith.

reverse for introducing some west into the course. As the given star or constellation rises higher in the sky, however, it becomes less useful by this method; at this time, one switches to another guide star that, by memory, one knows will rise at roughly a certain time, and in a certain place on the horizon. The relatively fixed North Star always provides a guide as well, as does the Southern Cross.

Using this method required not just these four markers, however, but a fully-memorized catalog of star rising times and positions, their transit time (how long they would be in the sky), and their setting time and position. Most of this information was archived in memory, rather than by using charts, although for instruction, navigation students did use starmaps made from sticks and shells throughout their years of demanding training. At sea, though, navigators relied on their memory. *Kavenga*, or *starpath navigation*, established a cultural encyclopedia of stars and constellation and used these natural guides for several thousand years of Polynesian navigation. Of course, the stars might not always be visible, so daytime methods were also devised, but it was the starpath that was at the heart of Polynesian navigation. None of the traditional methods used instruments, such as a magnetic compass, telescope, or graduated disc (e.g. a sextant). In this way, no material objects were required for this method.

Captain Cook, in common with other European explorers, regularly took Polynesian navigators aboard his own craft to help in his own navigation. Impressed by their abilities, he realized that their methods were sufficient to explain the entire peopling of the Pacific:

"In these *pahis* (voyaging ships), these people sail in those seas from Island to Island for several hundred Leagues (each league being c.3 miles), the Sun serving them for a compass by day and the Moon and stars by night. When this comes to be prov'd we Shall no longer be at a loss to know how the Islands lying in those Seas came to be people'd, for if the inhabitants of Uleitea have been at Islands laying 2 or 300 Leagues to the westward of them it cannot be doubted but that the inhabitants of those western Islands may have been at other seas far to westward of them and so we may trace them from Island to Island quite to the East Indies."[19]

In addition to using starpath navigation to identify position and direct a vessel, Polynesian navigators were expert at finding land. This was also done without equipment such as radar or telescope. For example, deep phosphorescence (*te lapa*) is known to occur from one foot to fathoms (6–12 feet) beneath the surface as streaks and moving glowing lights. Polynesian navigators observed that these often streaked in the direction of land from as far out as 80–100 miles out, but that they are noticeably absent when *closer* to land. In contrast, shallower luminescence (*te poura*) is known to increase much nearer the shore, though since it can be triggered by rain it was not always reliable. Drifting wood does not tell distance to land, but its condition (floating high, or waterlogged) can help indicate how long it has been afloat, and in different seasons different kinds of wood are more likely to be washed into the ocean. Volcanic activity – steam, ash and, at night, flashes on the horizon – could indicate land, and certain types of clouds, even seen from far off, were memorized as types that only occur (and hover over) islands. The activities of wildlife were also closely observed, of course, including birds known to go out to sea (flying high) in the morning but return (flying low) to land at night.[20]

Perhaps most astounding was the use of swell patterns to find land. A current running towards an island strikes the island, and to the trained eye, the swells that bounce back against the current's set were known to indicate the proximity of an island. Such swell patterns, and their interactions to produce complex, tapestry-like surfaces on the ocean, were memorized not just to find land but also to help to determine one's position. All of these methods were carefully memorized by student navigators over long years of expert training.

Some aids to navigation – in today's terminology ATON's – were devised and used by the Polynesians as well. 'Sighting stones', columns of rock on islands, were sometimes strategically placed, as in the case of 'The Stones for Voyaging' on Arorae island; these are 13 stones, each a flat coral slab 4x5 feet in size and 6 inches thick, set on edge and secured by paving so as to be visible from offshore

---

[19]   Cook (1777): 215.
[20]   See Ammarell (1999) and Feinberg (1988): 116. Note 4 in Lewis (1972): 3 states that the stability of Polynesia sailing vessels was an asset that made navigation easier.

and used to guide vessels away from Arorae – and towards other islands – at launch.[21]

For all of these reasons, Polynesian navigators were expert observers, often known by their bloodshot eyes.[22] But vision was not always at a premium; when David Lewis' guide, the navigator Tevake, became blind, he enlisted an apprentice to tell him what was happening in the sky, and he also began to sit in parts of the vessel where he could better feel the swells and complex water movements related to nearby islands. In this way, Lewis observed the incredible phenomenon of a blind navigator, expertly leading vessels on long journeys.

Using multiple methods and senses, then, Polynesian navigators became as integrated with their environments and vessels as jet pilots, when we invoke the concept of 'human–machine interface':

> "A Tahitian navigator [and his vessel] thus formed a single navigational device ... the navigator brought ancestral knowledge taught in the navigational schools together with the prow, masts, and rigging ... to follow the stars and find North and South, and [used] the hull as a swell-gauging instrument. Over long years ... the navigator learned to read the sea, stars and wind, until his knowledge became reflexive and embodied."[23]

Traditional Polynesian Sailing Voyages

Based on our knowledge of the winds and currents of the Pacific, and the sailing capacities of the Polynesian ships (many of which have been replicated and sailed for thousands of miles in the Pacific in the past 40 years), we can estimate that most voyages of exploration and discovery would have lasted less than 100 days. The ships themselves were clearly capable of such voyages. They were stocked with supplies designed for exploration: pre-cooked breadfruit, pastes of banana, coconuts, pounded taro root, baked fish (supplemented by fish caught at sea and cooked in a fire bowl fueled with coconut husks), dried breadfruit chips, various nuts and baked sweet potato, and all, we again know from experiment and the historical record, would keep at least for some months, and some indefinitely. Fresh water was carried in bamboo poles as well as pottery, and of course collected during rains. Other provisions included plants and animals for planting when land was discovered, such as hibiscus, sugar cane, fish-poisoning plants, Tahitian chestnut, various medicinal plants, bamboo roots, sago root, betel nut and others. Interestingly, many of these species must be protected from seawater, and some cannot be carried as seed but must be carried as shoots, being

---

[21] Lewis (1972): 316–317.
[22] Salmond (2008): 24.
[23] Salmond (2008): 30.

vegetatively-propagating plants.[24] Careful planning and proactive invention of storage technologies was mandatory.

In short, provisioning these ships for 30 days or more would not be difficult, and the vessels could easily last that long, giving them easy duration of a month at sea and certainly three times this if needed. Cruising at 7 knots for 24 hours a day is equivalent to travelling 168 nautical miles per day; if that is maintained for, let's say, 80 windy days in any given 100 days, voyagers could easily travel over 13,000 nautical miles – potentially bringing vast stretches of the Pacific islands into range of many others.

With regard to the generally easterly winds (that is, blowing from the east) into which Polynesian voyagers sailed, it has long been assumed that this movement would have been difficult. In terms of actual sailing, it is not difficult, but only somewhat laborious, and carried out by the same method of tacking as any European sailing vessel. Indeed, the easterly winds were considered an aid to travel east, because, if after a long voyage of exploration, nothing was found, it was easy to simply open the sails, turn the vessel's stern to the wind, and sail straight back to one's home island! This 'downwind return option' is used by sailors even today; arriving a bit upwind of a destination allows an easier downwind drift to pinpoint one's landing position.[25]

Finally, extensive experimental voyages using performance-accurate replicas of Polynesian double-hulled ships have, in the past 40 years, conclusively demonstrated that ancient vessels could sail wherever they wanted (in particular, the sailing conditions in the region of New Zealand absolutely prohibit its being reached by accidental drift from Samoa to the southwest). Some of the voyages completed include:[26]

- Samoa to Tahiti (W to E into wind) to evaluate windward capacity: 7 July – 21 August 1986.
- Raratonga (near Tahiti) to Aotearoa (New Zealand) to evaluate sailing into S seas: 21 Nov – 08 Dec 1985.
- Aotearoa (New Zealand) to Samoa (S to N) to evaluate 'return voyages': 1 May – 25 May 1986.
- Tahiti to Hawaii (S to N) to evaluate sailing into N seas: 24 April – 23 May 1987.
- Hawaii to Tahiti (N to S) to evaluate 'return voyages'; 10 Jul – 12 Aug 1985.

---

[24] Horridge (2008): 88.
[25] Finney (1994): 262–265.
[26] Finney (1994).

The Traditional Polynesian Voyaging Ethos of Exploration

Clearly the vessels and other technologies were available for ancient Polynesians to purposively explore the Pacific (Figure 6.2 shows excavation at an archaeological site where traces of Lapita people were discovered). But why did Polynesian voyagers set out, against the wind, in the first place? Anthropologist Ben Finney suggests the answer may be found in *exaption* rather than *adaptation*. Archaeologist Clive Gamble has argued that many early human migrations were driven more by exaption, the co-option of a previously-existing technical or cultural trait for a new purpose, than adaptation, the more explicit, novel adjustment of a certain trait to circumstances.[27] Finney provides an example in the form of a Fijian legend. Like all legends, it likely has a kernel of truth about the specific people mentioned, and like many Polynesian legends it certainly carries a significant message for us about the issue of motivations.

In this tale, chief Matai-welu arrived on an inhabited island whose chief allowed Matai-welu to explore for his own land; this exploration was only allowed, however, to windward, that is, towards the eastern side of the island. This Matai-welu and his sons did, tacking against the wind just as a European sailing vessel tacks against the wind by zigzagging. Matai-Welu and his sons eventually discovered and settled an uninhabited island, called Qele-levu. Later, when Matai-welu's sons matured, they sought their own fortunes and begged their father to be allowed 'to look for our own lands upwind from here'. Matai-welu gave that permission and the sons voyaged east into the wind, eventually finding and settling Vutuna.[28]

This interesting tale – we cannot be sure how old it is, but its hint at the Lapita's fighting east against the prevailing winds, which they certainly did – gives us a glimpse at some of the social issues that might have contributed towards colonization. To understand how, we need to know that in traditional Polynesian inheritance, the first son of a chief inherits everything; this means that sons later in succession have little to gain by waiting around for an inheritance; rather, they must move on to make their own fortunes. In this tale, presumably the sons referred to were not Matai-Welu's first son, but later sons, who set off east to make their own way in the Pacific. If so, this may be an example of the cultural rules of inheritance being *exapted* into repeated eastward explorations, as islands to the west were colonized and became the domains of powerful chiefs. For the later sons of chiefs, then, most of the time exploration east was a matter of survival, a way to ensure their own futures rather than simply an exercise in exploration.[29]

Another Polynesian legend, however, suggests that proactive exploration, simply for curiosity, was also instrumental. This is the legend of Ru, and his

---

[27] Gamble (1996).

[28] Biggs and Biggs (1975).

[29] Finney (1985): 173.

sister, Hina. Active voyagers, they ranged far across the oceans and eventually discovered every island in the Pacific. Somewhat at a loss for what to do next – thus belying the idea that exploration was only for self-interested gain – the pair noticed that the Moon, riding in the sky, was not simply a bright light; it appeared to have a surface, land that could be explored. The legend continues that Ru and Hina then built and outfitted a special voyaging ship to go to the Moon, and made the journey there.

This tale tells us of an exploration ethos – perhaps with several motivations – woven throughout traditional Polynesian culture. It was, sometimes more and sometimes less of course, a culture of explorers, guided by an ethos of extension, expansion, exploration and discover. *Ru and Hina* would be an appropriate name for a future vessel of space colonization.

Sometimes, pure exploration and even enjoyment do seem to have motivated Polynesian explorers. Today we classify the use of replicas of ancient sailing craft as adventures, and we exaggerate the dangers to sell the story. But just as surely as any Gloucester fisherman, traditional Polynesians were at ease with the ocean, and in some cases losing one's life at sea was considered proper, the 'sweet burial'. With this attitude, it is unsurprising to read report of 'ambitious' Polynesians voyagers, encountered by 19th-century European explorers, simply making adventures of "even 1000 miles, being not infrequently absent a year or two from home, wandering and gadding about from island to island … Samoa, Fiji, and all the Friendly Islands … [Tonga], Wallis, Fortuna …"[30]

Motivations, then, that we can detect in the history, legends and early European descriptions, so far include:[31]

- Chief's sons seeking to build their own chiefdoms.
- Increasing the land available for farming.
- The pride of navigators.
- Raiding and conquest/tribute payment and empire-building.
- Exile.
- Deep sea fishing (seeking larger fish offshore from inhabited islands).
- Trade.
- Discovery.

Oceanic Adaptive Lessons for Human Space Colonization

With regard to exaption and adaptation, as discussed above, it is probably not the case that *only* exaption *or* adaptation drove the exploration and colonization of the Pacific, but that it resulted from a combination of these (and other, unknown factors), sometimes the pendulum swinging more towards one

---

[30]   Dodd (1972): 278.
[31]   See Finney (1991).

motivator than another. What can we learn from this? What we can learn is that cultural variables were as involved in the successful colonization of the Pacific, as were technological variables. Indeed one might say that *despite* a relatively simple technology, a cultural engine drove the Pacific expansion. We have certainly seen before – in the case of 14th-century AD China – that simply possessing the technology for such endeavors does not guarantee them; cultural conditions might well prevent them. It is clear that this is the case today, in which we have the basic technologies to make, for example, a success of a Mars colony, but are not (apparently) equipped with a cultural motivator to make that happen. As we argue elsewhere in this book, sparking that cultural motivation will require a recasting of the entire concept of what we have termed 'humans-in-space', away from the negative connotations of technocracy, militarism, and commerce that so many associate with humans-in-space as it has been presented in the last half century. In this light, the case of Ru and Hina should remind us of the significance of the cultural element in beginning the extraterrestrial adaptation.

This cultural point is important. Many NASA managers have lamented that the 'can-do' culture that put humans on the Moon was replaced by a management-top-heavy, 'can't-do' culture that focused more on limits than possibilities (discussed in the next chapter). The vital, crucial culture of innovation decayed away. While NASA, we argue, will not be at the forefront of human space colonization – it will be largely orchestrated by private industry – any organizations involved will be well-served by developing the intelligences and expertise to make a success of human space colonization. And this building of creative cultures that actually believe they can achieve human space adaptation can be intelligently guided by taking the lessons we have learned from an evolutionarily-informed, adaptive approach to the key features of intelligence and creativity themselves, discussed in Chapter 5. For all of these reasons, the lessons of the Pacific colonization are most significant to the building of adaptive cultures that will support human space colonization.

7 Building an Adaptive Framework for Human Space Colonization

> *"Man has risen, not fallen. He can choose to develop his capacities as the highest animal and try to rise still farther, or he can choose to do otherwise. The choice is his responsibility, and his alone. There is no automatism that will carry him upward without choice or effort and there is no trend solely in the right direction. Evolution has no purpose; man must supply this for himself."*
>
> George Gaylord Simpson[1]

> *"An organism is more than a bundle of separate adaptations; it is a coordinated complex of adaptations. The adaptation of a hawk for [predation] involves the combination of several features: soaring flight, telescopic vision, sharp grasping talons, strong body, and hooked tearing beak."*
>
> Vere Grant[2]

We have seen, now, that our species is one of many. We have had a successful, two-million year run so far. But to persist as a species we cannot rest on our laurels. We must continue to use humanity's unique capacity for proactive adaptation – so precious in comparison with the reactive nature of adaptation in every other life form we know of – to survive. This includes, we argue, the colonization of space. Inspired by our past successes, such as the glories of Pacific and Arctic colonization, and the human space program to date, we now consider how humanity will orchestrate the fourth major adaptation, which we call the Extraterrestrial Adaptation. A project as vast as the expansion of our species' geographical range to include off-Earth environments will require more creative proaction than anything yet attempted. Because this will be an adaptive endeavor, we must shape it with everything we know about evolution, the author of adaptation.

In this chapter we describe some elements of an adaptive framework for human space colonization. Such a framework would guide space colonization with both broad concepts, such as that of adaptation to space, rather than the

---

[1]   Simpson (1949): 310.
[2]   Grant (1963): 115.

conquest of space, and very specific concepts such as those at the foundations of evolutionary computing and the new field of biomimicry, which builds technology according to, as one author puts it, '3.8 billion years of field testing.'

We show how up-to-date conceptions of evolution, adaptation, proaction, intelligence and technology can be assembled into an evolutionarily-informed structure that can guide the planning and carrying out of human space colonization.

A Closer Look at Adaptation

So far in this book we have argued that since human space colonization will be a continuation of natural human evolution, and we have specified that it would be, essentially, adaptation – the fitting of a life form to an environment in which its ancestors could not survive. We have seen some examples of human and other adaptations. With evolution already outlined, and some examples of human adaptation laid out, it is time to more closely examine adaptation as a guide to the building of an evolutionary, life-based approach to human space colonization. In this chapter, we begin with an examination of adaptation and then go on to see what lessons can be learned from it that can be used to shape human approaches to the colonization of space, both materially and philosophically.

Defining Adaptation

In 1919, naturalist Frank Sumner examined the history of the idea of adaptation, quoting an early adopter of the concept as writing that life is the "continuous adjustment of internal relations to external relations."[3] This highlights the root of the concept of adaptation in the concept of the maintenance of equilibrium in living things, which is a good start. Sumner then went on to comprehensively dismantle the idea that adaptations were consciously-invented designs, adding to the then-rising tide of Darwinian evolutionism (Darwin's *On the Origin of Species* had only been published in November 1859) and the abandonment of ideas that we now classify as *intelligent design*. The importance here is general, but profound, as it aided in the consideration of life forms from a natural (rather than supernatural) perspective, setting the basis on which, today, all we know of adaptation is grounded.

As we saw earlier in this book, the term adaptation can be used as a noun, in respect of a particular characteristic that makes life possible in a given environment, or in terms of the process by which such characteristics accumulate in an organism's genome over time. Biologist C.L. Prosser carefully

---

[3]   Sumner (1919): 23.

examines the concept of adaptation, and provides a definition that is one of many in the literature, but quite well suited to the project of human space colonization:

> "The capacity to live in an environment not occupied by forebears indicates that adaptive evolution has occurred. The essence of evolution is the production and replication of adaptive diversity."[4]

Prosser also indicates that in biological systems the genetic heritage of a lineage might restrict the range of variations that might eventually become adaptation and that adaptations are environmentally induced. The lessons here are that while these determinants of adaptation apply to non-human life forms, for humanity things are significantly different. Respectively, the human invention of adaptations – both behavioral and technological – allows the overcoming of biological heritage (to a large degree), and humanity's capacity for niche construction makes it possible for humanity to create niches rather than simply react to environments. Of course, we also carry a cognitive heritage, and that we should be careful not to be bound by outdated conceptual frameworks and we should be careful to monitor our own paradigms for fit with reality. The phenomenon of *normalization of deviance*, in which complex cultures internalize and accept failures as acceptable, until the day they become catastrophic, can be avoided by applying lessons from this aspect of adaptation.

Learning from Adaptation for Human Space Colonization

Evolutionary approaches to a variety of human goals beyond biology have been attempted, sometimes with success, over the last few decades. For example, the field of biomimicry, which we introduced earlier in this book, explicitly seeks engineering solutions from naturally-evolved life forms. Another example is seen in evolutionary computing, which is based on the evolutionary principles of mutation and selection to arrive at solutions. In this approach, the power of mutation is harnessed by having a computer program 'build' thousands of designs of a given solution, each with a slight variation, and the power of selection is harnessed by having the program evaluate each solution for its closeness to the performance wished as the end product, selecting those variations of design that approach the desired performance as templates for the next round of mutations, and deleting those that do not achieve some performance criteria.[5] Figure 7.1 shows a traditionally-designed girder built with 'common sense' compared to a much stronger design (with an unlikely appearance to the 'common sense' eye) derived from an evolutionary algorithm. Another example includes the attempt to learn from adaptive principles in shaping responses to climate change.

---

[4]   Prosser (1986): 2.
[5]   Eiben and Smith (2008).

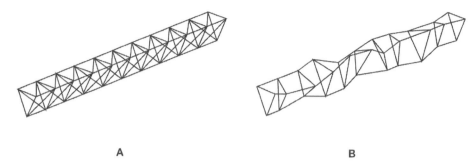

A   B

Figure 7.1. Girders Designed by Traditional (A) and Evolutionary (B) Methods. Image by Cameron M. Smith after Figure 1.7 of Eiben and Smith (1998).

In the same way, in this section we discuss some of the lessons of adaptation, as studied by palaeontologists, biologists and evolutionists, as they could assist in the building of an adaptive framework for human space colonization. A review of several decades of literature on adaptation reveals at least three lessons of significance.[6]

Lesson 1: Adaptations Arise from Ecological Opportunities

A prerequisite for adaptive radiation, according to evolutionary biologists Jonathan B. Losos and D. Luke Mahler, is the availability of an ecological niche, or habitat, for colonization.[7] For human space colonization, the lesson is deceptively simple: begin by identifying habitats available for human colonization. But this becomes quite complex when we consider that such niches might themselves change over time, requiring humanity to adjust, or that human intentions and wishes might themselves change over time, such that the niche is no longer sufficient to sustain human populations. Therefore, an adaptive approach to identifying suitable niches for human colonization should include consideration of the time dimension. This requires consideration of time units; years of human life are familiar, but other units, such as generations (concerning demographics and population genetics) or units related to agricultural potential and productivity, and many others, should be considered. Such complexity reminds us that we should begin such an adaptive approach now.

---

[6]   Our review spanned dozens of books and papers on adaptation, theoretically, generally and specifically; references are found where appropriate in the text.

[7]   Losos and Mahler (2010).

Lesson 2: Successful Adaptations in Nature Can Be Characterized and Learned From

Palaeobiologist and MacArthur Fellow Geerat J. Vermeij has studied the characteristics of adaptive systems at large, and specific adaptive properties of individual organisms, for over four decades. Based on this expertise, gained from an understanding of both individual organism anatomy and patterns of adaptation in the fossil record over vast time periods, in a 2008 paper he discussed a number of general methods that life forms, at various scales, use in "... averting, neutralizing, blunting or eliminating unpredictable threats."[8] We reiterate these below as they could apply to human space colonization, and then go on to discuss their adaptive corollaries.

1. *Tolerance via passive resistance.* This method is, as we will mention again below, generally unsuitable for such a proactive species as *Homo sapiens*.
2. *Active engagement to disable and/or eliminate a threat with force.* This method requires a good knowledge of a given threat, and structure dedicated, individually or collaboratively, to carry out adaptation.
3. *Increase knowledge of threat to make it more predictable.* Clearly, knowing of a threat, and being able to predict its frequency in time, location in space, and magnitude of effects, is of great importance.
4. *Increase unpredictable behavior to prevent threat from predicting your actions.* This is modifying one's own behavior so as to become temporally and/or spatially unavailable to the given threat.
5. *Isolate and starve threat of resources.* This is unlikely to be particularly useful for human space colonization issues except in the case of combating communicable disease, which will be a major concern while founding off-Earth populations are rather small, as they will likely be at first.
6. *Use redundancy/modularity to make local losses sustainable.* As we will discuss below, this is an important lesson, already familiar to engineers, for example, who often build towards redundancy and distribution of effort such that no single failure will result in systemic failure.
7. *Decentralize control to make local losses sustainable.* This addresses modularity, essentially, of control and decision-making, for the same essential reason as mentioned in #6.

Additionally, Vermeij suggests a number of general issues to keep in mind when coping with threats, as policy suggestions – some of which are rough corollaries of the points made above – based on "... major facts revealed by the fossil record of life and by the principles governing the ways in which living entities interact and evolve";[9] we reiterate these below, with slight modifications

---

[8] Vermeij (2008): 25–41.
[9] Vermeij (2008): 40–41.

that do not change the content, but only reorder it somewhat for our concerns with human space colonization.

1. *Threats from unpredictable challenges rarely entirely go away, and no adaptation is ever 100% effective.* It is clear that in nature, every adaptation has its costs.[10] Applied to human space colonization, this is simply a lesson in maturity: we cannot expect our technological or other adaptive solutions to off-Earth environments to be 100% effective; naturally, we want them to be as effective as possible, but because we ourselves change, and the Universe changes, and because of error resulting from misunderstanding conditions to which one is adapting, manufacturing flaws, and even malice, it is to be expected that, as in other human adaptive endeavors, there will be problems. We must be willing to accept and overcome those problems. This is probably not the most pressing issue in designing effective human space colonization, as the people involved will necessarily be optimistic and future-oriented. Today, we accept (in the US) roughly 100 automobile fatalities every day, for a total of over 45,000 in 2005;[11] we also accept that while the chances of being in a fatal aviation crash are low, they remain, and always will; and these fatalities are accepted in daily life, much less in a larger, future-oriented goal of human space colonization. Of course, we are not saying that space colonization should be cavalier to danger, only that it will include inherent dangers. Again, for the people involved, however, this does not seem to be a deterrent.

2. *Adaptations cost, reducing resources available for other functions.* The lesson here is that we must expect to pay for human space colonization, and the currencies will include both capital wealth as well as cultural stability. Regarding culture, in particular, we human space colonists will have to expect cultural change rather than resist it. In short, the idea of replicating Earth-based culture in space colonies will have to be left behind. Off-Earth colonies will not be much like anything humanity has built before.

3. *Passive resistance can support a long life-span, but only in a small ecological niche.* Passive resistance – such as hibernation in some animals or transformation to a spore state in some microbes – does not seem to be particularly available to humanity as an adaptive mechanism at present. As we have shown throughout this book, human adaptation endeavors so far have been largely active rather than passive. Still, this approach could help in technological design, in which solutions might be found that require minimal resources. One example is the passive thermal control program used in the Apollo Moon missions, in which the spacecraft, in

---

[10]   For some examples, see Wallace and Srb (1964): 101–107.
[11]   Nauman et al. (2010).

transit to and from the Moon, regulated temperature by slowly rotating the vessel during flight such that no single surface was 'cooked' by sunlight of 'frozen' in shade. Another rather passive approach to potential threats to self is seen in blue-water sailing, in which long voyages that might appear terrifically dangerous on the surface can be planned such that areas of bad conditions and seasons of bad conditions can be entirely avoided by careful scheduling of the voyage.[12] This approach is echoed, regarding technical solutions, in the phrase *sometimes the best airplane is no airplane.*[13]

4. *Actively meeting threats has a very high cost, demanding a constant supply of resources; this option should always be supplemented with other means of adaptation if it is to be sustained.* The lesson here is that for any given adaptive goal, multiple responses should be designed so that, ideally, no single solution carries an inordinate burden, and if any such element of the adaptation fails, its failure might be absorbed or passed on to other solutions. The essential lesson here is of great importance: the significance of distributed effort and modularity. As we have seen elsewhere in this book, adaptations are often modular and interconnected. Because organisms are composed of many individual (though interrelated) traits, and the biological heritage of a given organism is not entirely reshaped by evolutionary processes in the short term, adaptations at large can be seen as modular. In terms of human design of adaptations for certain environments, this should promote adaptations designed to be modular, rather than monolithic, meaning that we may ask, during design, what other elements of our adaptation, must, should, or could usefully be addressed by this element of this adaptation?

5. *The greatest adaptive potential is found in flexible systems with weak centralized control, high modularity, localization, and both unique and redundant elements.* This is perhaps one of the most important of Vermeij's lessons, and it is reiterated by other studies of long-term adaptation in living systems:

> "... no capability should be invested in single units, or depend on single components of infrastructure in a system. Diffuse responsibility prevents errors made by one party from crippling the entire system and allows for the testing of competing adaptive hypotheses ... Error ... can be reduced in its frequency and its effects, particularly in a system characterized by modularity and redundancy, but it cannot be eliminated."[14]

We will return to this point below, and here it is sufficient to say that

---

[12]   Vigor (2001).
[13]   Altschuller (1998).
[14]   Vermeij (2008): 40–41.

these principles – flexibility, modularity and distributed rather than centralized control – should be considered in all human space colonization plans, from engineering design of individual solutions, settlement layout design, and longer-term issues such as culture change.

6. *Predicting a threat, and acting to prevent it from arising, is more cost effective than fighting it if it arises.* At least one lesson that can be derived from this is that understanding the environments we wish to colonize will, of course, require good knowledge and characterizations of them as guides to solutions – but it should be remembered that such characterizations can 'fossilize', ignoring the fact that conditions can change. To accommodate such change, plans of all kinds should include contingencies for cases of change, appropriate to the expected – and sometimes predictable – magnitude and character of changes in space environments. In this way, space weather forecasting, understanding of seasonal changes in various off-Earth environments and other predictable and unpredictable (but known) variables will be important elements of an adaptive approach to human space colonization.

7. *Over both long and short time, adaptive measures improve with use.* This reminds us that the sooner we begin to design an adaptive framework for human space colonization, the sooner we will improve on our design, even as a result of simple studies and simulations. For example, years of experience accumulated by members of the Mars Society's simulated Mars habitats – in the US and Canada, with plans for another in Iceland – have begun to build such experience. The lessons from these simulations will be both concrete, as in the design of Mars habitat technologies, as well as social and cultural. We should remember, however, that these model early, exploration-mode human presence on Mars, and not communities *per se.*

Some general adaptive lessons to be derived from Vermeij's insights have been mentioned, and certainly an adaptive approach to planning and carrying out human space colonization could be positively influenced by these lessons. It is too early to specify exactly how, but it is difficult to imagine that this knowledge should go unused.

Lesson 3: Benefits of Collaboration

One of the discoveries resulting from the last three decades of the 'genomic revolution' that has allowed us to understand, in great, unfolding detail, the molecular bases of evolution is that no life form is solitary. Even the microbes, which we might think of as solitary, anonymous 'germs', have active lives involving signaling and receipt of signals from other microbes.[15] Biomimicry

---

[15]   West et al. (2007).

pioneer Janine Benyus, whom we will mention again later in this chapter, has outlined some of the adaptive lessons of collaboration – intentional and unintentional – in living communities. First, collaborative efforts allow the pooling of resources, as when many sensors are used in the same search for resources, rather than a single sensor. Second, collaborative efforts can 'extend' the body in space and time, as in the case of ants that form bridges for other ants to crawl across. Also, collaborative effort can allow rotation of effort among community members, such that some rest while others work. Collaborations also allow the 'swapping' of skills, such that reciprocal arrangements maximize the potential of differences rather than uniformity in a system. Individual risk is also minimized by collaboration, and habitats may be divided among community members in a way that each is fitted to their own special niche, maintaining diversity rather than trying to 'enforce' uniformity, which has multiple downsides.

As in some other cases, it is beyond our scope in this book to suggest how such collaboration benefits should materially inform human space colonization plans, but it seems that only a severely lacking imagination would be unable to benefit from this knowledge.

Lesson 4: Biological Adaptations are Sustainable

Life forms that overrun their resources experience population collapse or extinction, a point being largely resolutely ignored by modern civilization. This will not work, of course, in off-Earth colonies, where resources will be scarce (relative to on Earth). When planning solutions selective pressures, then, it will be necessary to consider sustainability. This is not just jumping on the sustainability bandwagon – it is recognition of the short-term decision-making that guides most of our lives, and that we must move away from them. Essayist George Monbiot summed up the situation this way:

> "We live in a dream world. With a small, rational part of the brain, we recognize that our existence is governed by material realities, and that, as those realities change, so will our lives. But underlying this awareness is the deep semi-consciousness that absorbs the moment in which we live, then generalizes it, projecting our future lives as repeated instances of the present. This, not the superficial world of our reason, is our true reality."[16]

Monbiot points out that we act more on short-term concerns than long-term, and the result has of course been a great many 'solutions' to pressures that, in the long run, have resulted in the many unsustainable 'solutions' we labor under today. Their cost should be calculated to include the cost of changing to

---

[16] Monbiot (2003).

sustainable solutions later on. Clearly, it will be better to design human space adaptations to be sustainable. Luckily there is a great trend towards sustainable practices today, in fields as diverse as the management of interpersonal relationships to engineering. A significant element of sustainable practice is to realistically model the change of conditions over time. This (and other considerations) can aid in the prediction of discernible changes in either the local conditions, or the existing adaptation itself (a behavior or a technology, for example) that would requiring further adaptive adjustment.[17] Knowing what these 'red flags' are, allows subtle and flexible adaptations. Generally speaking, such awareness promotes a solutions philosophy that internalizes the nature of the Universe itself: change. Whenever our adaptations can be designed to be continuously adaptive, they will be improved.

A shortlist, then, of guiding adaptive principles to inform human space colonization plans should include the following:

- Sustainability.
- Modularity.
- Intelligent integration.
- Collaboration.
- Decision-making decentralization.

Of course, while these aspects of biological adaptation might well improve our adaptations to space environments, we do not need to be enslaved by them; they can be thought of as guidelines, not shackles. But inasmuch as various kinds of design endeavors have been profitably guided by evolutionary principles, awareness of these elements of billions of years of adaptations in Earth's living systems should better align human adaptations to the dynamic – rather than static – nature of the Universe, and should be of interest and use to humanity.

A Framework for Human Space Colonization

To build these principles into an 'adaptive framework for human space colonization', we need to define what that is. It is a cognitive scaffold, the structure of which conditions the thoughts that lead to actions. An example from the emerging field of biomimicry – engineering design fundamentally informed by evolution – is illustrative. First outlined by Janine Benyus this list of general observations structures the progress of her field:[18]

1. Nature runs on sunlight.
2. Natures uses only the energy it needs.
3. Nature fits form to function.

---

[17]   Alland mentions this in relation to the monitoring of human populations (1970): 37.
[18]   Benyus (1997): 3.

4. Nature recycles everything.
5. Nature rewards cooperation.
6. Nature banks on diversity.
7. Nature demands local expertise.
8. Nature curbs excesses from within.
9. Nature taps the power of limits.

Benyus' principles were, in turn, adapted by architects Daniel Vallero and Chris Brasier in their principles of sustainable, evolutionarily-informed design:[19]

- Waste nothing (less is more).
- Adapt to the place (learn from indigenous strategies).
- Use free resources (renweable and abundant resources).
- Optimize rather than maximize (use synergies, reduce mechanization).
- Create a livable environment (build for life, not against life).

As mentioned, such principles are not meant to be shackles, but rather a guiding hand.

If human space colonization is to succeed, we argue it will have to be designed as a long-term, adaptive endeavor, proactively guided to realize the Extra-terrestrial Adaptation. To do this, it cannot be carried out with the short-term, goal-seeking behavior that has characterized the human space program so far. Our entire way of thinking of human space colonization must be revolutionized such that nature and evolution, rather than being fought by technocratic civilization, are *adapted to*.

Precisely how the current 'humans-in-space' mindset inhibits productive thought about human space colonization, we will come to later. For the moment, we can ask: how does one overcome embedded structures of thought? Essentially, a *paradigm shift* must be made. The paradigm, in scientific theory as formulated by Thomas Kuhn, is a set of accepted practices, a consensus on what is to be observed and analyzed; it conditions the questions that are asked and how the results of experiments are analyzed. When it allows errors, Kuhn continues, the paradigm becomes harmful to the generation of new knowledge, and a paradigm shift is required. Resistance to paradigm shift is often found in the phenomenon of *groupthink*, in which a collective set of systematic errors is perpetuated by a group because of general resistance to change and the collective momentum of the status-quo, can be difficult to overcome.[20] It is exactly this kind of status-quo about humans-in-space, engrained in the popular culture, that must be overcome.

The good news is that human culture, while capable of tremendous entrenchment, is also capable of tremendous flexibility and change. In Western civilization in the last few centuries we have seen significant tidal shifts related

---

[19] This list is adapted from Vallero and Brasier (2008): 26.
[20] See Janis (1982) and, for paradigm shifts, Kuhn (1962).

to the growth of scientific knowledge and the development of Enlightenment principles. We have shifted from a conception of a limited, human-centered Universe to an infinite Universe in which humanity is one of countless life forms, and in politics we have moved from monarchy to democracy. In religion many have moved from authoritarian Catholicism to independent Protestantism, and tolerance for a general multiplicity of religious views – and more recently, even atheism – has flourished, even in the founding documents of a number of nations. These are significant changes in the history of Western civilization, arguably more sweeping and rapid than any change in any previous civilization, such as the Maya, Inca, Sumerian or Egyptian.

Current attempts to understand the kind of broad cultural change represented by what we envision as a paradigm-level change in humanity's conception of humans-in-space is often carried out by anthropologists as well as marketers:

> "There is an emerging paradigm called 'slow design' which represents a series of diverse opportunities to encourage consumption in which people reflect on what really nourishes them while simultaneously reducing negative environmental, social and economic impacts. Like all paradigm shifts, however small or large, it subtly or robustly challenges the status quo and involves new ways of seeing, knowing, and believing. It demands that 'design' explores a new philosophy and purpose. As a paradigm challenge, it has some parallels to that between the eighteenth-century Newtonian rationalist, reductionist and (now) traditional view of science and the holistic science of Goethe."[21]

The environmental movement that arose after the 1959–1973 Space Race, partially driven by the images of Earth suspended dramatically in lifeless space, is a more recent paradigm-level shift towards more conscious use of resources than ever before. Such conceptual shifts begin with education; they are encoded in textbooks. In Vallero and Brasier's recent undergraduate text, *Sustainable Design: The Science of Sustainability and Green Engineering*, we see an excellent example of the thin end of a paradigm shift wedge, working at the level of basic college education for tomorrow's architects and urban planners:

> "It is only through creating a better understanding of the natural world that new strategies can emerge to replace the entrenched design mind-sets that have relied on traditional schemes steeped in an exploitation of nature. Designs of much of the past four centuries have assumed an almost inexhaustible supply of resources. We have ignored basic thermodynamics."[22]

The sustainability paradigm, indeed, is even being implemented in the field of engineering, at similar educational levels. The US National Academy of

---

[21] Fuad-Luke (2010): 134.
[22] Vallero and Brasier (2008): 2.

Engineering's new publication, *The Engineer of 2020: Visions of Engineering in the New Century* includes this statement:

> "It is our aspiration that engineers will continue to be leaders in the movement toward the use of wise, informed, and economically sustainable development. This should begin in our educational institutions and be founded in the basic tenets of the engineering profession and its actions."[23]

Building a new cognitive framework, then, that will condition basic thought about the larger thought-domain of humans-in-space, will best be done by a series of cognitive shifts implemented by its inclusion in undergraduate (and earlier) education.

Cognitive Shifts Necessary to Promote Human Space Colonization

Before more explicitly describing an evolutionary/adaptive framework to shape human space colonization, we describe below some of the more general cognitive shifts required to promote human space colonization as an evolutionary, human-centered adaptive effort. These shifts cannot be made quickly or precisely, and the sooner we begin making them – often simply by thinking carefully about how we talk about human space colonization – the better.

Cold War/Militaristic → Humanistic/Families in Space/People

This shift involves demilitarizing the motives and carrying out of space access and colonization. While military agencies have served an important role in developing space access so far, space colonization will be in large part about civilian organizations, families and communities building a civilian rather than military culture off-Earth. Additionally, the cold-war overtones of control and competition related to the 'Space Race Era' must be replaced with language that emphasizes everyday civilian life.

This shift will occur somewhat naturally as private industry is increasingly engaged in providing direct access to space, and as legal mechanisms establish codes of conduct in off-Earth environments. The essential message here is the humanization of space for peaceful purposes.

Industrial/Mechanical → Human/Organic

This shift involves moving away from conceiving of science, technology, mechanism and industry as ends in themselves – or necessarily linked, as in

---

23 National Academy of Engineering (2010): 50–51.

the name of Oregon's 'Oregon Museum of Science and Industry' – and towards conceiving of these as human inventions meant to aid in adaptation. Science, for example, is a method, and can be used for industry or the pure generation of knowledge, or even for enjoyment, and need not always be associated with its material precipitates and their inevitable commodification. On *Discovery* magazine's main website, for example (and in many newspapers), as on Fox News' main website, news from science and technology are unnecessarily (and, strictly speaking, improperly) combined in sections titled 'Science/Technology'.

This large-scale cultural shift of values would recognize that the science that makes us smart and adaptive is not necessarily bound to technology and commerce. Further, it would emphasize that technology is not necessarily an end, but a tool for adaptation. Working with a high technology–human interface in highly demanding environments, Special Operations Command General Schoonmaker recently pointed out that what is important is to equip men, not to man equipment.

'The Right Stuff'/NASA Complex/Professional → Everyman/Pioneer

This shift involves moving away from the conception of humans-in-space as the sole domain of a small, privileged few highly-trained space professionals, and towards the inclusion of 'everyman', the general public, because it will be large populations of civilian non-space professionals that will form the bulk of off-Earth colonists. While some high-level professionals, such as spacecraft pilots, will always of course be necessary, space and access to space must be disengaged from the concept of exclusivity.

Implementing this shift will, like some others, occur naturally as NASA is increasingly devoted to long-range space exploration and research, and public access to space becomes cheaper and more commonplace. As in many cases, a substantial part of the shift takes place simply in conversation, socialization and education. We would do well to point out that any given space colonist will more likely be a farmer than a test pilot.

Short-term, Limited Goals → Long-Term, Infinite Future

This shift involves moving away from the conception of humans-in-space as a short-term 'project' with an end-point, towards a conception of a long-term future. In other words, space colonization should be recast as a *beginning* for the rest of human history, than any kind of pinnacle or end. There is no knowing what will come of humans colonizing off-Earth environments, but humanity remaining on Earth will, inevitably, result in extinction in the long term; the Sun, as we know, will in 5 billion years or so burn out, incinerating the Earth. For this reason, the author H.G. Wells noted, even at the turn of the 19th–20th century, that for humanity "It is the Universe, or nothing."

This cognitive shift requires an emphasis on time and evolution and their internalization by humans; they must be believed as being important. Deep time

– the recognition that the Earth is billions, rather than some tens of thousand of years old – must be internalized, as must evolution. There is no end to evolution except in extinction, and the ultimate goal of human space colonization is the prevention of extinction of *Homo sapiens sapiens*, at least for some millions of years, at which point we may well have evolved into another variety of life that is not, hopefully, Earthbound, but distributed widely in the Solar System and perhaps further.

Exploration → Colonization

While humanity will always be exploring, the role of space colonization must be separated from the current exploration-mode conceptions of humans-in-space. Most space colonists will be family people, building new civilizations, rather than the professional explorers noted above. In time, they will not leave Earth laden with the idea that they must return to Earth.

Achieving this shift will occur naturally as off-Earth colonies are established and flourish, achieving a sort of cultural normality; 'Of *course* there is a Mars colony' we hope people will argue in the near future, rather than 'Why would we build a Mars colony?' But even in the run-up towards the time of the first off-Earth colonies, the shift will have to take place to convince the general public that space is not just a place where exploration takes place, and where only explorers go; it must be recast as a place where common people can go to have families and a future.

Conquest of Nature → Adaptation to Nature

This shift involves a significant reconsideration of the relationship of humanity to the rest of the Universe: away from the concept that humanity does or can 'conquer' Nature (including space) and towards the concept that humanity adapts to Nature, and will adapt to space environments. In the 1950s, Werner von Braun's influential *Conquest of Space* series, published in a thrillingly-illustrated series of *Collier's* magazine issues, was important to building support for manned space exploration in the 20th century. However, the conception of humanity conquering Nature is at odds with everything we know about evolution. There is, in fact, no Nature to conquer – only environments to which humanity can try to adapt; and if those environments change, humanity will have to change with them. There is no such thing as conquest, in the long term. In evolutionary time, one must remain flexible and continue to adapt. 'Conquest' is specific, and therefore limiting. Adaptability is general and limitless.

This shift, as others noted above, also requires learning and internalizing the fact that humanity is one of millions of species, and has only relatively recently arisen (and this by the evolutionary process of adaptation), and could easily become extinct. Our 'conquest' of any given environment has not yet stood a long test of evolutionary time. Therefore, science education, and general speech and socialization should put a premium on the concept of adaptation rather than conquest.

Limitations to Humanity → Increasing Human Options

This shift involves a move away from concepts of human space colonization as being a province of hardship, stress, and limits to freedom, and towards a conception of its *removing* limitations and providing humanity with many more options – in an infinite Universe – than are available when constrained to living on Earth. A popular conception of the perceived limits to life off-Earth is found in an essay by William Hodges:

> "... for generations [of early space colonists] the settlers will be poor. They will spend their entire lives in conditions of extreme material deprivation and high risk of death."[24]

While the material life of early space colonists will certainly be different from those of, say, modern suburban Americans, 'deprivation' seems a strong term. In another essay (in the same volume in which Hodges' essay appears), Edward Regis, Jr. counters that, because off-Earth human colonies will have to be well-managed and will be unable to support the kind of explosive population growth we have seen on Earth:

> "Far from being severe deprivation ... all the generations to be born ... are to be well provided for throughout their life spans, something that could not be said for most of the children who have been born on Earth."[25]

Achieving this shift requires engaging the mind with a creative, expansive, and ultimately optimistic philosophy when discussing the human future in space. Certainly, space colonies will for a long time require different modes of life than many of us live on Earth. But it is important to remember that many humans today – the bulk, actually – live in conditions of poverty and overcrowding. Such conditions will not be possible at least in the earlier generations of off-Earth migration, as resources will be too limited to allow them. Artwork depicting prospective off-Earth settlements must emphasize such optimism and expanse, rather than constraints.

Barely Possible → Possible and Routine

This shift requires a move away from the conception of humans-in-space as an exception, something that is rarely done because of its expense and technical

---

[24]   Hodges (1985): 135.
[25]   Regis (1985): 255.

challenges, towards a conception of humans routinely going into – and remaining in – off-Earth (space) places.

Like some other shifts, this will occur rather naturally, as more people enter space and spend more time there. Ultimately, children will be born off-Earth, and never visit Earth. They will be the first extraterrestrial human generation. Such a provocative concept must be communicated and explored, again, with optimism and a de-emphasis of the machinery in exchange for an emphasis on the humans who will build and use that machinery.

Destiny/March of Progress → Human-Directed Evolution

This shift requires a move away from the conception of humanity being destined to march triumphantly into space, and towards the conception that humanity will only succeed in space colonization through proaction. As the evolutionist G.G. Simpson was quoted in the opening to this chapter, *"There is no automatism that will carry [humanity] upward without choice or effort and there is no trend solely in the right direction. Evolution has no purpose; man must supply this for himself."*

This shift, like others mentioned above, will require an internalization of the fact that evolution is not internally driven 'to' create any given form of life; nor does evolution usher any given form of life through predetermined stages. Evolution, as explained earlier in this book, is simply the change of life forms over time, a *consequence* of the facts of replication, variation and selection. When evolution is properly understood, then, as must be achieved through education and in the general public culture, it is clear that humanity cannot wait for or simply expect preordained success in colonizing space, but will have to build that success by proactive adaptation.

Costly Luxury → Responsible Investment

This shift requires a move away from the idea that human-in-space activities are a costly luxury towards seeing such – including human space colonization – as a responsible investment in the human future. Economically, there is perhaps never a 'good' time to engage in space exploration (not to mention colonization); space exploration has been expensive to date (though human access to space will in the near future become radically cheaper), and startup costs for human space colonization will certainly be high in dollar terms. Recently a writer for the *London Times'* 'Eureka' supplement stated:

> "For a child of the Seventies like me, spoon-fed on Apollo Moon landings, this seems a sorry state of affairs. Glorious pictures from distant worlds are all very well, but after a while they begin to feel like brochures for luxury holidays the West can no longer afford."[26]

---

[26] Miller (2011): 58.

This shift can be assisted by pointing out that while cost in dollars will be high at first, our species routinely invests in long-term projects at vast expense, such as the US 2011 defense budget, officially stated at nearly $700 billion. Robert Zubrin's 2001 cost estimate for establishing a first colony on Mars, keep in mind, was $400 billion, a cost considered too high by US. Congress at that time. However the costs are calculated and modified, it is clear (as mentioned in an earlier chapter in this book) that, particularly when distributed among several nations and spread over some years, space colonization would be of high-to-moderate cost, but is not impossible. In fact, it should be pointed out that avoiding space colonization might be expensive now but pay off in the end, as cost incurred now could prevent bankruptcy – extinction of our species, or the beginning of a new Dark Age by any number of factors.

Spinoff Mentality → Survival Mentality

This shift, related to the one mentioned immediately above, involves a move away from the question 'what will we on Earth gain from space colonization?' towards the realization that what we will gain is survival – or at least a better chance for the survival of our species than we would have if we remain on Earth. NASA has for most of its existence published lists of *spinoff* inventions – microwave ovens, digital wristwatches and so on, resulting from space-exploration technologies – in part to justify its existence. But such material gains, while substantial, cannot be the justification for humans-in-space activities, including human space colonization. The mentality must, at least, include the survival of our species as a significant and worthwhile result of colonizing space.

This shift will require a greater appreciation, in the public culture, of the threats to human civilization, and our species, posed by ourselves (e.g. warfare), Earth-based biology (e.g. plague) and extraterrestrial threats (e.g. Earth being struck by a comet or asteroid). Every few years the media run dramatic stories about solar flares that threaten communication satellites, and, thus, internet access and even cell phone operation. Improving an awareness of the possible threats from outside Earth could be accomplished by, among other methods, publishing 'space weather' bulletins in the mass media, in tandem with the usual Earth weather forecasts.

Technospeak→ Common Speech

This shift, related to all of the above in one way or another, entails a move away from the specialized, technocratic language of acronyms and unfamiliar terms relating to humans-in-space and towards a 'normalization' of how we speak about, and think about, humans-in-space.

Certainly, acronyms and specialized language are necessary and appropriate for managing space programs, but they are alienating and counter-productive when attempting to engage the general public, beyond a certain point. Having said this, 'everyday people' can internalize a sense of grandeur and even feel proud of humanity's achieving the technical expertise required to put humans into space by using technical terms when thinking or talking about space exploration. But the interface between engineers and the general public must be shaped so that it does not alienate, but rather opens the field of thought for discussion. Engineers and other scientists interfacing with the public should receive some training in popularizing their field (effective use of metaphor, for example), and basic education should include discussion of science as neutral, something that can be used responsibly (electric lights) or abused (nuclear and biological weaponry).

This shift is subtle, but, as in poetry, its subtlety is derived from its essential simplicity. Using different words – some of which have not yet been invented – to communicate to one another about humans-in-space at large can help to normalize the concept of human space colonization. In *Realspace: the Fate of Physical Presence in the Digital Age, On and Off Planet,* P. Levinson recently wrote:

> "So the need is not to find a new medium of communications, but a better way of communicating about space through media already in play. We can begin with the very way we talk about our space vehicles. Starships are more enticing than space shuttles; cloud offices are more inviting than space stations. A name such as 'Mermaid Dune' – given to a Martian rock site identified in NASA's 1997 Pathfinder mission – is a minor step in the right direction. It humanizes an alien environment, in contrast to 98-BLG-35 and the alphanumeric soup of appellations often given to new meteorites, asteroids and even possible planets discovered in other solar systems."[27]

Words, in short, are important. We are happy to eat *escargot*, or *Rocky mountain oysters*, but few Westerners will be tempted to eat, respectively, snails, or sheep testicles. Clearly, we cannot 'talk our way into space', but just as clearly, words are important. In Table 7.1 we present some ideas in this significant realm of *renaming*.

Principles of an Evolutionary/Adaptive Framework for Human Space Colonization

The cognitive shifts noted above would largely place humans and evolution at the center of human space colonization, moving away from technocracy. We can now turn to just what would entail an evolutionary and adaptive framework for human space colonization.

Generally, human space colonization should be considered in terms of

---

27   Levinson (2003): 12.

Table 7.1. Old and New Terminology Relating to Humans-in-Space. These simple suggestions indicate the kind of terminological changes that could help to humanize the idea of human space colonization.

Old Words/Terms	New Words/Terms
space exploration	space travel
astronaut	space traveler, colonist, voyageur
capsule, spaceship, rocket	vessel, ship, pirogue, lighter, caravel, spore, pod, seed, tendril, galleon, bergantin
base/habitat/colony	village, home, city (use Earth terms)
extraterrestrial	off-Earth

survival value for our species, and therefore be informed by everything we know about evolutionary survival. Specifically, three lessons from palaeobiology should be built into the cognitive framework that shapes human space colonization.

The first lesson is that the longest-lived species exist in *large populations* allowing for survivors of even widespread species disasters. Humanity currently has a population of about seven billion, a number that has doubled in just the last 40 years. In terms of sheer numbers our species can be considered successful, but we should recall that it is in the invisible world of the microbe that the most numerous life forms are found. One major implication of the large-population lesson is that when considering human space colonization, we should move away from ideas of small, fragile, outpost populations and towards the idea of humans moving to space not to live marginally, but in profusion. Space colonization should be designed to promote large populations.

The second lesson is that long-surviving populations are *widespread* in both general space (e.g. across the entire planet) and specific ecological niches (e.g. tigers being capable of surviving both hot, tropical environments and cold, temperate environments). Such distribution also allows for survival of some of the species when others are selected against either in the short- or long-term. One lesson from this is that space colonization should be designed to occur not just in Earth-orbiting communities, or Moon colonies, or on the surface of Mars and other nearby planets and bodies such as asteroids, but in *all* of these locations and more, extending out, eventually, to other planetary systems via interstellar exploration and migration. Further, space colonization should be designed not as a concretely-delineated project with a discrete end goal – as was the Apollo Moon landing program – but as an open-ended, infinite process with the nebulous goal of spreading humanity far and wide, eventually to even the interstellar realm.

The third specific lesson of evolutionary survival is that long-lived species have *behavioral and/or genetic diversity*. Genetic diversity is the lifeblood of healthy gene pools because as selective environments change – as they are bound to do because the natural world is in constant flux – selective pressures change, and variations which were yesterday of little value become necessary for survival under new selective pressures. The same applies to behavior, especially for a species (such as humanity) that relies more on its behavior than its genetic constitution to survive. While our biological structure must allow basic survival, we humans shield it from many selective pressures with myriad behavioral adaptations, including the construction of tools, shelters, clothing and methods of water purification, cooking, and so on, extending even to the immensely-important complexities of human social interaction. Just as geneticists measure the health of a given species in terms of its genetic variability, then, human cultural health should be considered in terms of our cultural diversity, which allows the behavioral flexibility required for long-term survival. Of course, precisely this diversity has been used by tyrants, for as long as we know, as the basis of conflict, and as with any invention or process, it must be admitted that it can be the source of either solutions or problems. The chief lesson here is that space colonization should be planned to incorporate both genetic and cultural variation.

Values

The values, then, of an evolutionarily-informed, adaptive framework for human space colonization must fundamentally include the main characteristic of evolution: change itself. Because the Universe – from the largest to smallest scales – changes, evolution and life are characterized by change, not fixity. Accordingly, while a framework itself is meant to guide thought and action, in our case it must not constrain thought, but guide it to embrace the universal reality of change.

To be informed by what we know of evolution at large, and human evolution and adaptation, such a framework should be designed with the following lessons in mind:

- The Universe belongs to all humanity.
- Access to space and the Universe is a human right, not a privilege of the few.
- The colonization of space will be a natural continuation of human adaptation.
- All designs and plans should be informed by the 3.8 billion years of evolution.
- The Universe beyond space will provide freedom, not limits, to humanity.

To weave these core concepts into a framework for successful human space colonization will result in a very different character of considering humans-in-space than we carry today in popular culture. Human space colonization, informed with the principles and cognitive shifts so far noted, will be (and will be perceived as) a more organic, natural, evolutionary, adaptive, humanistic, possible and reasonable action. The framework itself will be moved away from its technocratic, unnatural, exclusionist and expensive associations widespread today.

With the general zeitgeist changed in this way, other means of building an adaptive framework become much more conceivable. Cognitively, new thoughtscapes are cleared. We identify and discuss several below.

Build Towards Evolutionary Success

As mentioned, our species requires three characteristics (at least) to ensure long-term, evolutionary success:

- A large population.
- A widespread population.
- A general rather than specialized behavioral and genetic repertoire.

Ensuring these can be achieved by human space colonization, which will ultimately increase the human population, distribute it in space and time, and both capitalize on and stimulate behavioral and genetic diversity, which are the measures of the health of our species.

Use Adaptive Principles

The principle of adaptation is *adjustment to prevailing circumstances*. Therefore, we must change with Nature, not fight the unconquerable.

Adaptive systems are normally modular and lack central control. Plans, execution and even building materials and components should be designed with these principles in mind. Also, adaptation results in *branching* events, in which A gives rise to B, which is somewhat different from, but recognizably derived from, A. This obtains for culture, technology and biology. For evolutionary biologists, new functional complexes are required to count as evolutionary novelties. When branching events – divergences resulting from innovations – occur in human space colonization, they should be considered in their own terms rather than as mutations or aberrations.

Use the Potential of Exaption

Exaption, as mentioned earlier in this book, is adaptation by using not entirely novel structures, but structures that once served another purpose. Evolutionists commonly use the term *co-option* to identify gene functions related to exaption, for example when specific types of hair became elaborated and evolved into feathers. At least one lesson is that a design or procedure that has no appreciable function today may be of great adaptive value tomorrow; another is that rather than reinvent the wheel when it comes to, say, the colonization of Mars, we might well learn from – and adapt methods used by – colonists from our own prehistory, including lessons from their social arrangements, architecture, and even methods of farming certain crops. The former lesson leads directly into the fields of invention and creativity – humanity's ace-up-the-sleeve – which we will discuss later in this chapter.

Remember Proaction

The distinction of human evolution is that it can be guided and rapid. No other life form even knows that it is evolving, or what it means that selective pressure X has recently changed and will have Y effects in n generations. But humans can know all of this, as we have seen throughout this book, and we proactively tailor our evolution; in fact we can change our own selective environments by the staggeringly potent phenomenon of niche construction. The lesson is that proaction requires intelligent direction, and that the better our plans, which depend on high-quality understanding of the cosmos and its properties, including the life-building, life-preserving process of evolution, the better will be our adaptations. One ramification of this principle is that knowledge and science should be (as discussed elsewhere in this chapter) disassociated from industry and commerce because while a given discovery might serve neither of those fields, it might well serve to improve our proactive adaptations.

Use of Nature-Based & Evolutionary Technologies

Worldwide over the past three decades engineers have found that evolutionary mechanisms have over long time and wide space already arrived at a great many 'engineering solutions' of use to humans. For example, Benoit Mandelbrot's identification of the fractal geometry of nature was used directly to improve cell phone antennas, all of which today use fractal design to increase the antenna efficiency by 'packing' large antennas into small spaces. Indeed, many of the forthcoming generations of robotic exploration vehicles are being designed on the templates of insects, snakes and other animals. Also, genetic algorithm computing – which cranks out millions of slightly-varying designs, and selects against those that do not meet the designer's specifications – has made significant strides in the past two decades. Ultimately, starships might look like fishes, or clusters of frog eggs; they might respirate or pulse, they might be grown rather than built. There is no more active a field in engineering today than the previously-mentioned *biomimicry*, which seeks engineering solutions for human problems in the world of living things already built by billions of years of natural selection. The field of evolutionary computing, which uses algorithms that produce many slightly-varying results that are evaluated by a selector programmed with certain parameters in mind, is also exciting and promising as it harnesses the power of evolution.

Use Technology for Humanity/Build Better Information Ecologies

It is easy to let technology take the upper hand, as we see and feel in modern society where we are routinely frustrated by poorly-designed technologies. The task to improve technology for human space colonization is to make technology our servant, not our master. This is already underway throughout many engineering and design fields, with a focus on ergonomics and better human–

technology interfaces. For example, there is a movement underway to completely redesign aircraft cockpit instrumentation – which remains in many ways bound to designs from the 1930s, using difficult-to-read dials and gauges – with new displays that are enormously more readable, reducing cockpit task stress and increasing pilot performance. Some are even currently experimenting with cockpit signaling methods that would involve many of a pilot's very sensitive senses – motion, touch, hearing, and so on – to communicate states of a given craft in a way that is more bodily than external to the body. Again, this capitalizes on the already highly-sensitive system of the human frame, built and tested by evolution over a long period.

Generate Cognitive Variation and Creativity by Encouraging Do-It-Yourself Projects

Space access and humans-in-space are already being shifted from governmental authority to private enterprise. This demystification and opening-up process is extremely important in that it opens humans-in-space concept to a much wider and, importantly, *participatory* population. NASA has long held design contests at universities and high schools and even elementary schools, recognizing the enormous potential of distributed, parallel thinking; more brains at work means more concepts and more chances of finding workable concepts. The 'do-it-yourself' mentality also engages the general public, making the project of human space colonization their own, rather than something delivered to them with all the usual strings attached that come with government delivery. The X-Prize has, of course, already worked, leading to the production of craft capable of suborbital space access. Such programs do not need, always, to award money: Britain's 2011 Foundation has in place a non-monetary award – which may eventually be as coveted as a Nobel prize or a military medal – for the first people to make an expedition to the south and north poles of Mars.

It is important to remind the general public that everyone can participate in the colonization of space in one way or another. Figure 7.2, for example, shows a home-built pressure suit being developed by Cameron M. Smith for balloon exploration of the lower stratosphere. This project is meant to demystify the technology of 'space adaptation', highlighting the idea that Mars colonists, for example, will not be able to summon a technician every time they need adjustment to such technologies, but will have to engage in this work themselves, requiring the development of cheap, reliable and easily-maintainable equipment for living off of Earth.

Continue Pure-Knowledge Research in Biology, Adaptation and Extinction, and Learn From Them

While exaption (discussed above) is important, proactive adaptation is as well, and there is no telling what creative solutions for human space colonization will be stimulated by the discovery of adaptations in ancient life forms by geneticists,

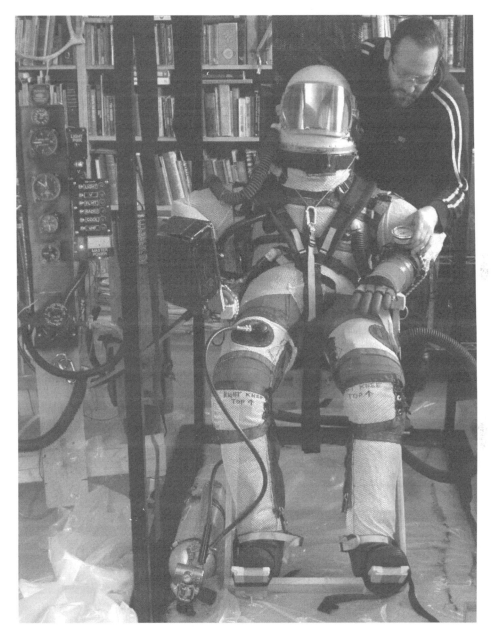

Figure 7.2. Home-Built Pressure Suit Developed by Cameron M. Smith. Photo by Cameron M. Smith.

palaeontologists and palaeobiologists; in this book we have suggested a mere handful in the hope of opening a new topic. Of course, identifying and characterizing life forms on Mars, or one of the Solar System's moons, could enormously aid our understanding of adaptation and evolution, and these efforts should continue while human space colonization is being planned and carried out.

The significance of biology – even microbiology – for making human space colonization successful cannot be overstated. After nearly 200 years, biology is far from being 'finished'; in fact, according to many it is currently being revolutionized by the last 30 years of genomic studies, which have led to an "incomparably more complex vision" of the fundamental processes of evolution. Chief among the new concepts of evolutionary biology are the *coevolution* of species – leading to better understanding of mutualisms and synergies used to inform the new engineering field of biomimicry – and Horizontal Gene Transfer, the phenomenon whereby species A is now known to acquire, during the life course, genetic material from species B. This newly-appreciated fact forces not abandonment of Darwinian biology, but a substantial inclusion of Lamarckian 'inheritance of acquired characteristics' in a new Darwinism. Considering this, re-examination of adaptive processes already known, in light of an emerging new evolutionary paradigm, will certainly give us new insight in how adaptation occurs, and that will increase our ability to use evolutionary biology to inform human space colonization via proactive adaptation.

Of course, to make the lessons of extinction and adaptation as revealed by biology useful to the human space colonization effort, it cannot simply be hoped that the information will move from biology to human space colonization planners. What is learned from biology must be translated into the language of human space colonization and professionally applied to it. For this reason HSC planning should include a structure for liaison with biology (and anthropology). Such a liaison could be applied in university courses at, for example, the International Space University in Strasbourg, France.

Grow More Effective Human Space Colonization Decision Cultures

It is well-known that the two space shuttle disasters were ultimately caused by poor management, characterized by middle-level managers being unresponsive to lower-level engineers, whose warnings about dangers therefore never reached higher-level program directors with 'Go/No Go' authority. Therefore it is not at all trivial or 'soft' to work at improving the problem-solving communities that will develop and carry out human space colonization activities. Two main fields could be of great help.

First, cultural anthropology is itself the study of human social organizations, and it has a 150-year archive of experience in documenting, and – more recently – attempting to explain the myriad 'ways to be human'. This information could be very helpful in building an HSC culture that does not look like any management structure we have seen before, and in fact it should be unique because its function is unique: not space exploration in short-term missions nor robotic exploration of distant bodies, but long-term migration of human populations from Earth to extraterrestrial habitats. Only anthropology has the expertise to address such a project in the wider scheme.

The second field is less obvious but potentially just as important. Recent biological investigations have shown that horizontal gene transfer, the phenomenon by which species 'adopt' DNA from other, contemporaneous species not through reproduction but simply by absorbing free-floating DNA from their surroundings – is common in bacteria and not uncommon among multicellular life forms, and therefore Lamarck's 'inheritance of acquired characteristics' conception of evolution was not as mistaken as we normally believe. The significance of this for understanding human culture is enormous because it has long been appreciated that human cultural information transmission is Lamarckian: humans are enculturated not by the gene, but by what they pick up from their phenotypic environment, the 'inheritance of acquired characteristics' during the course of life and passing that inherited (learned) material on to both their peers and the next generation. Just as evolutionary models of cultural information transmission have been strongly structured by (but not a slave to) traditional population genetics, to update our concepts of culture and culture change we need to appreciate the lessons (sill unknown at this early stage) of Horizontal Gene Transfer for culture. We need, then, another investigation of the parallels between cultural and biological evolution, but this time informed by the new understanding of evolution resulting from three decades of genomics.

Finally, archaeologist Charles Redman has reviewed the literature on cultural fit and misfit to environment, with an eye to understanding the failure of the Viking colonies in Greenland mentioned earlier in this book. He concludes that:

> "First, the model used in making a decision might be based on another ecosystem that had surface similarities, but critical differences. Second, insufficient detail of information might be presented, leading one to overgeneralize the problem. Third, there may only be a short observational series, because it is a truly new situation where there is little realistic experience. Fourth, the managers might feel detached, being socially and geographically distant from the producers. Fifth, the reaction of the managers might be out of phase with the problem; in other words, too little, too late. Sixth, the managers may perceive the potential problem, but not feel obligated to take action: 'it's someone else's problem'."[28]

---

[28] Redman (2001):226.

Certainly, to build better decision cultures for human space colonization, we might well learn from other such analyses of human adaptive failures in the modern as well as pre-modern world.

Work to Human Strengths, and Beware Human Weaknesses

Human strengths and weaknesses have been generally revealed by evolutionary and anthropological studies tracking how we have gone well, and poorly, as a species. Some of these lessons are described below:

Strengths

- Individuality and behavioral flexibility.
- Creativity and rapid problem-solving.
- Subtle and powerful language and communication, based on symbolism and metaphor.
- Ability to reason in the abstract.
- Technical capacity to invent solutions to existing or perceived obstacles.
- Ability to overcome impulses.

Weaknesses

- Categorization function of mind can ignore differences and lead to stereotyping.
- Historical context of individual can significantly condition cognitive options.
- Short-term, small-space bubble of episodic awareness.
- Status-quo rationalization.
- Susceptibility to emotional appeals by self-aggrandizers.
- Immune system issues.
- Adversity to 'failure'.

Regarding the final point above, the 20th-century genius-inventor Buckminster Fuller forcefully held that the word 'failure' should be eliminated from our vocabulary; and engineer Henry Petroski recently commented that:

> "Indeed, failures appear to be inevitable in the wake of prolonged success, which encourages lower margins of safety. Failures in turn lead to greater safety margins, and, hence, new periods of success. To understand what engineering is and what engineers do is to understand how failures can happen and how they can contribute more than successes to advance technology."[29]

---

[29]　Petroski (1985): xii.

Failure, then, in human space colonization – and it sure to occur at one or all levels at some time or another – should be avoided, of course, but considered differently than it is as present when it does happen. How we learn from failure should be reconsidered as well.

Synthesis

Rather than superficially jumping on the evolutionary/biomimicry/sustainability design bandwagons, we have suggested significant benefits that will attend the casting of our conception of human space colonization in terms of evolution and adaptation. More concretely, an evolutionary/adaptive framework for human space colonization recognizes and promotes the following ideas as guides for our thoughts and actions.

An **evolutionary approach recognizes that the Universe changes**, and that one cannot fight nature, at least not for long. In engineering and design, for example, engineer C. Dyke discusses the 'entropy bill' that, no matter what, must be paid, and must be considered from first principles even in the growth of complex systems as large as cities:

> "Finally, we can think of an extreme situation ecologists often encounter: a species adapting itself to one dimension of its environment, rigidly committing itself to maximal efficiency within that environment at the expense of the capacity to react effectively to environmental change. Such species are always in grave danger of being evolutionary dead ends, because environmental change destroys the narrowly circumscribed conditions for their survival ... Similarly, we can think of the 'one product' city: for instance, the mining town. It is extremely vulnerable to changes in, let us say, the economic climate. Organized rigidly for one purpose as it is, if the lode dries up, or the market changes, etc., it will take enormous effort to reorganize for success in the new environment. *Here is where the entropy debt is paid!* ... All the while that the single resource city is functioning successfully, its economic accounts are satisfactorily balanced ... The entropy debt that is piling up is hidden until the sustaining flux is interrupted ... In short, capitalism and its ideology are and have always been ahistorical. And if we think history – as any serious structural change – can be prevented, we can be comfortable with equilibrium bookkeeping. Evolutionary theory, on the other hand ... has to be historical ... Equilibrium bookkeeping will not work in evolutionary biology ... [and] it will not work in ecology; and if it will not work in ecology, then eventually it will not work in economics either. For the second law of thermodynamics insists that we cannot stop ecological change, and we cannot seal ourselves off from the changes except temporarily and at extremely high cost ... Within standard economics the decisions are all framable as cost/

benefit decisions. The costs and benefits are cast as resource allocations ... We *never,* except in the most superficial ways, examine the relationships between our patterns of social organization and the rate of material flow needed to sustain them. This optimal efficiency is never satisfactorily connected to the time dependent thermodynamic conditions that really matter."[30]

Dynamism and change, then, the hallmarks of evolution and the Universe at large, must be considered at every level of thought and design regarding human space colonization. Following this, we find that **adaptation – the ability to maintain or increase fitness during change – is measured as mutability**. Unless we wish to fight change and entropy, we must remember that:

"Evolution is change, adaptation to new circumstances. Evolutionists see a world of process, of flux, incomplete, imperfectly known."[31]

To be in line with nature, then, mutability and plasticity are the keys to survival. For humans this means maintaining genetic and cultural variation, not fixity. This does not mean that there is no role for well-tested routine; the wide adoption of checklists by aviation crews in the last three decades has led to significant safety improvements. At the same time, the ability to react and work 'outside the box', to innovate and experiment, must be valued. A dramatic case is seen in the recent 'miracle on the Hudson', where the pilot who successfully landed the stricken jetliner did not do so according to any pre-programmed procedure or checklist; there was no time for these. Rather, Captain Sullenberger attributes his safe landing to his long history of piloting sailplanes, unpowered aircraft that carry only bare-bones instrumentation and are flown almost entirely by 'feel' and innovative decision-making that occurs from moment to moment in a continuous flow.

In an important parallel, evolutionary and anthropological studies have also shown that evolution is not internally driven in a unilineal, progressive sense. While many species do indeed become better-suited to their environments over time, that is a result of chance, not an internal engine; in fact most species *do not* accomplish adaptation, and in the long term become extinct, as have 99% of all life forms that have ever existed on Earth. Thus, we humans must cast off our self-conception as an impervious pinnacle of evolution and recognize that we are but yet another of billions of manifestations of life energy on Earth through time. While our unique consciousness allows us to alter our own evolution, we must not be fooled into thinking that we were destined to survival; we might well become extinct, but that fate can be significantly reduced by increasing our numbers, expanding our population off of Earth, and maintaining our genetic

---

[30] Dyke (1988): 363–365.
[31] Kehoe (1985): 181.

Figure 7.3. An Early Primate and the Stars. Only by proactive adaptation will this line of life persist beyond the life of the Earth. Painting courtesy of Mr. Michael Rubino.

and cultural diversity. Another lesson resulting from recognizing that we humans were not inevitable is that we have to make our own future. Sitting back in wait for inevitable progress would most likely be disastrous. Only by proaction has our species survived to date, and only by continued proaction will we persist.

That necessary proaction, in an evolutionary sense, recognizes that our species' trump card is the intelligence that has buffered us, with inventions, from selective pressures. **Significantly, intelligence is today largely defined by creativity and adaptability, not rigid 'fact memory'**. Here we see a marrying of intelligence, adaptation, evolution and the noted dynamism of the Universe and the selective pressures that shape our species. 'Smart' systems are adaptive, changeable and sustainable, while 'dumb' systems have no proactive or reactive capacity, are fixed rather than changeable, and wasteful, leading to a finite life. Human space colonies and communities – on moons, in orbiting structures, in multigenerational interstellar ships, or what have you – will succeed only if they bank on our species' capacity for intelligence, which translates into a capacity for proactive adaptation at a speed and specificity not seen in any other life form.

Considering this, **our species' best likelihood for maximizing success in space colonization will be to promote, in education and culture at large, creativity**. Once considered the realm of fuzzy, pop-psychology, the study of creativity – which is the ability to generate novel thought associations – is today at the heart of the study of intelligence. How creative solutions – intelligence – is promoted and generated is also widely studied. It is turning out that while memory and recall capacity are of course important for short-term goals and procedures, in the larger picture intelligent innovation is more facilitated by experiences that blur the concrete boundaries of ideas and allow new associations of ideas to build into new realities. The seat of intelligence, then, is in novel associations rather than in routines. To overcome a certain ossification of routines in industrial diving, for example, in 2008 a symposium on the future of diving technologies at Trondheim, Norway, included a jazz musician to help foster new thinking about diving. The musician, Bjorn Alterhaug, writes:

> "Creativity is understood as a slow-floating improvisation, a process based on knowledge, proficiency, and reflection over time ... Improvisation may be described as immediate action, where there is no time for reflections in the moment, yet action based on deep internalized knowledge, proficiency and creativity. Improvisation is the creativity manifested in the moment's invention."[32]

Further, F.J. Barrett, in a paper titled *Creativity and Improvisation in Jazz and Organizations* , indicates that seven methods used in jazz improvisation can assist in building organizations that actively promote creativity:[33]

1. Provocative competence: deliberate efforts to interrupt habits.
2. Embracing errors as sources of learning.
3. Shared orientation towards minimal structures that allow maximum flexibility.
4. Task distribution: negotiation and dialogue for dynamic synchronization.
5. Valuing retrospection.
6. 'Hanging out'; building a community of practice.
7. Taking turns in soloing and supporting/accompanying.

Are such methods trivial, human-resources management fads? If so, they should not be. Exhaustive analysis of catastrophic failures at NASA, such as the losses of the *Columbia* and *Challenger*, reveal that they were significantly the result of inflexible, uncreative, rigid management cultures of both NASA and some of its sub-contractors. These 'can-do' cultures, for all their expertise and previous creativity, lost their creativity and became unreactive, resistant to the change that characterizes the Universe. This is clearly evidenced in the

---

[32] Altgerhaug (2009): 162.
[33] This list is adapted from Barrett (2000).

frustration conveyed by a memo written by a subcontractor engineer investigating the properties of the failed booster O-rings that led to loss of shuttle *Columbia* on 21 January 1986:

> "We are currently being hog-tied by paperwork every time we try to accomplish anything. I understand that for production programs, the paperwork is necessary. However, for a priority, short schedule investigation, it makes accomplishment of our goals in a timely manner extremely difficult, if not impossible. We need the authority to bypass some of the paperwork jungle."[34]

In her widely-acclaimed analysis of the management culture that was on watch for both shuttle disasters, Vaughan concluded that while risks attended every space flight, and not all should be laid at the doorstep of management when dealing with cutting-edge, high-risk technology, the management culture that existed was overly-complex (tragically, largely a result of contractor negotiations rather than design optimization). In short, while a complex management culture was inevitable because of the scale and complexity of the project, that culture did not have to be (to use her term) *bureaupathological*:

> "The cause of the [1986] disaster was a mistake embedded in the banality of organizational life and facilitated by an environment of scarcity and competition, an unprecedented, uncertain technology, incrementalism ... routinization, organizational and interorganizational structures, and a complex culture."[35]

This excursion into the world of intelligence, creativity, innovation, and what promotes them is not meant to criticize unproductively, but to inform our future organizations and plans, which will be a result of our intelligence, with an evolutionary perspective. That perspective tells us that the expertise we need to adapt to non-Earth environments should hold flexibility at a premium; this lesson is derived not from a human resource management fad, but from the profound lessons of biology and evolution. Those lessons can shape better expertise. A recent study of expert performance indicates that expertise has four commonly-understood features, all of which can be improved by improvisation:[36]

- Deep conceptual understanding: not just fact memory, but facts as parts of a complex conceptual framework.
- Integrated knowledge: facts are connected with others, and with systems; they are not independent.
- Adaptive expertise: in addition to standards and procedures used in the

---

[34] The memo is reproduced in Vaughan (1996): 453.
[35] Vaughan (1996): xiv.
[36] This list is adapted from Sawyer (2008).

appropriate circumstances, experts are able to improvise new solutions and do so enthusiastically rather than reluctantly.

- Collaborative skills: experts are comfortable with and engage in collaborations that are not necessarily hierarchically arranged, but arranged by talent.

Clearly, understanding and refining our understanding of what it means to be intelligent, to be adaptive, can only increase the likelihood of success of human space colonization, and should be incorporated into every level of education as well as the framework for human space colonization. Despite the enlightening ideas just discussed, W.B. Arthur of the Santa Fe institute recently wrote, in *The Nature of Technology: What it Is and How it Evolves,* about our understanding of *inventiveness* itself:

> "We are really asking how invention happens. And, strangely, given its importance, there is no satisfying answer to this in modern thinking about technology ... today invention occupies a place in technology like that of 'mind' or 'consciousness' in psychology; people are willing to talk about it but not really to explain what it is. Textbooks mention it, but hurry past it quickly to avoid explaining how it works."[37]

Later in the book Arthur indicates his belief that innovation is about **mental association** of previously-unassociated ideas, and visualization of outcomes: a process of "linking a need with a principle (some generic use of an effect) that will satisfy it". Thus, "To invent something is to find it in what previously exists."

A final, broad lesson from evolutionary studies, related to the points just made about intelligence, refers to the development and even our conceptions of both science and technology. In an evolutionary sense, technology – from the simplest stone tool to the Large Hadron Collider – is clearly the tool of humanity, not an end. **Thus, to avoid technocracy, technology must be thought of with more subtlety than has characterized the technocratic atmosphere of the last century, which has often put human concerns second to our inventions**. Regarding this point, technologist M.I. Detzouros writes:

> "Tomorrow's computers, appliances and mobile gadgets, together with the Information Marketplace formed by billions of interconnected people and devices, can do great things for us ... but not the way we're going! The important quest ... is to make our information systems human-centered ... [For example, consider] the human servitude fault ... where your telephone says, "*If you want marketing,*

---

37   Arthur (2006): 189.

press 1., if you want engineering, press 2., et cetera." Here you are, a noble human being, at the tentacle of a $50.00 computer, obediently executing machine-level instructions."[38]

A first step towards implementing the necessary realignment of our relationship with technology is a subtler understanding of the scientific method that leads to technology. In *The Golem at Large: What You Should Know About Technology*, H. Collins and T. Pinch write:

> "The personality of science is neither that of a chivalrous knight nor pitiless juggernaut. What, then, is science? Science is a golem. A golem is a creature of Jewish mythology. It is a humanoid made by man from clay and water ... It is powerful. It grows a little more powerful every day. It will follow orders, do your work ... but it is clumsy [and] without control a golem may destroy its masters with its flailing vigour ... it is a little daft. A golem cannot be blamed if it is doing its best. But we must not expect too much. A golem ... is the creature of our art and our craft."[39]

Concluding on technology (rather than science), the authors continue:

> "... a better metaphor for technology is the cook or gardener. No one who is clumsy in the kitchen or the garden shed can possibly doubt there are those who are more expert than themselves. Yet no one expects ... a perfect souffle or herbaceous border every time ... Take away the mystery and the fundamentalism and we can see frontier technology as the application of expertise in trying circumstances. That is how we must find our way through the technological world."

More directly, technologist W.B. Arthur concludes in *The Nature of Technology: What it Is and How it Evolves* that we must master our inventions, not serve them, and that while any invention could separate humans from challenge, meaning, purpose and alignment with nature:

> "Where technology separates us from these things it brings a type of death. But where it enhances these, it affirms life. It affirms our humanness."[40]

The lesson is that to improve the science that leads to the technology that preserves our species – from antibiotics to pressure suits – a more creative, flexible, and life-sciences-informed path is needed. Technology and science must be humanized. Since science and technology – but not rigid technocracy – will be key to successful human space colonization, it is in everyone's interest to

---

[38] Dertouzos (2002): 183–184.
[39] Collins and Pinch (1998): 1.
[40] Arthur (2006): 215–216.

reconsider even these foundations in a more meaningful way in the building of our evolutionary and adaptive framework for human space colonization. Many calls for subtler, less-linear, more evolutionarily-informed approaches to science that can better teach humanity about the living world are currently underway – perhaps inevitably considering the exponential accumulation of knowledge in the past century – even in basic biology. In this respect, microbiologist Carl Woese has recently written:

> "A society that permits biology to become an engineering discipline, that allows science to slip into the role of changing the living world without trying to understand it, is a danger to itself. Modern society knows that it desperately needs to learn how to live in harmony with the biosphere. Today more than ever we are in need of a science of biology that helps us to do this, shows the way. An engineering biology might still show how to get there; it just doesn't know where "there" is."[41]

The result of this kind of humanization of science and technology could only improve what we have referred to broadly as the *humanization of space*. We argue that that humanization will be necessary not only to sharpen our thinking about human space colonization, but also to improve the chances of carrying it out successfully, both in the long-term and in large space. The actual planning and carrying out of human space colonization should be informed by the principles we have outlined in this chapter. They will lead to a long-term, sustainable, organic adaptation to space for humanity, rather than a short-term and ultimately failed attempt to 'conquer space'.

---

[41] Woese (2004): 173.

8 Distant Lands Unknown: Informed Speculation on the Human Future in Space

"As soon as somebody demonstrates the art of flying, settlers from our species of man will not be lacking ... Given ships or sails adapted to the breezes of heaven, there will be those who will not shrink even from that vast expanse."

Johannes Kepler, in a letter to Galileo Galiliei, 1610[1]

"One of the gladdest moments in human life me thinks, is the departure upon a distant journey into unknown lands."

Sir Richard Francis Burton[2]

Throughout this book we have seen that life normally spreads into any habitat it can endure, sometimes shaping that environment passively, and, more characteristic of humanity, sometimes shaping environments consciously and proactively. We have been successful because we have been explorers. If instinct is a biologically-based compulsion, somehow hardwired, one might say that our species' inclination towards exploration is instinctual. But it seems to be relatively new, rather than primitive; our closest living relatives, the chimps and gorillas, are not such active explorers. Our remarkable zest for ranging out and discovery date to the last few million years, when our genus, *Homo*, began to actively expand out of Africa. On this timescale, we are the descendants of about 60,000, 30-year generations of explorers, for many of whom it was natural to wonder what lay over the horizon, and to satisfy that curiosity through exploration. And in the last few decades our exploration has even extended from the surface of the Earth, putting human beings on the Moon, and our robotic probes even beyond the edge of our Solar System.

The question today is whether, as a species, humanity will continue to explore beyond the Earth, and eventually colonize other places in our Solar System, and, later, even farther out from the cradle. Will we once again move beyond what is known, relying as we have for millions of years on our tools, our capacity to learn and adapt?

---

[1] See *Conversation with the Sidereal Messenger.*
[2] Burton (1872): 16.

At this writing, in late 2011, the human space programs of the United States and the Russian Federation are very limited compared to what they were in the past few decades. The dreams of the 1960s and 1970s – further exploration of the Moon and colonization of Mars, and even the construction of orbital space colonies – have been abandoned. The international space station is a small orbiting scientific outpost that depends entirely on the Earth for its support and supplies, and it has been pointed out that while for many years it could have been experimenting with how to be a self-sustaining colony – growing its own food, for instance – short-minded concerns kept it tethered to Earth in such respects, the loss of a tremendous learning opportunity. Many will ask how one can even consider the colonization of space at such a time. But, we argue, we do not stop producing art, or writing literature, because economic times are bad, and in fact it might be at the worst economic times that expansive, positive ideas are most important.

Although in the long term it seems certain that humanity will move from its cradle, we cannot predict when will and resources will be assembled to begin a human adaption to off-Earth environments. We can only say a few words about where humanity might first go, and what our species will have to do to survive there.

Where Can We Go? The Moon

Practically speaking, there are only two immediately obvious places for humanity to begin adaptation to off-Earth environments: the Moon, and Mars. The last trip to the Moon was made in 1972, and today our image of the Moon is strongly conditioned by our exploration-mode exposure to it in the form of small teams (pairs of men) exploring the surface for less than a week, and returning with samples to be analyzed. We have no conception of the Moon as an actual habitat. Could humanity live there? We know that, at base, humanity needs a certain amount of water, food, breathing gas, and enough control of atmospheric pressure and temperature to survive. Regarding water, studies since the 1970s have shown that significant quantities of water ice exist in certain areas on the Moon, particularly in the northern and southern polar regions; one NASA estimate is that over 6 billion tons of water ice – about 750 million gallons – of water ice exists in these areas.[3] With various methods, this water ice could be melted as drinking water (which could be recycled) and used for agriculture, and it could also serve as a source of breathing oxygen. Regarding food, plants could be grown using this water and hydroponic (soil-free) methods, and this would be necessary as the lunar regolith (soil) is poor in nitrogen.[4] Regarding breathing gas, splitting water ice to derive breathing oxygen would be relatively

---

[3]   See http://nssdc.gsfc.nasa.gov/planetary/ice/ice_moon.html.
[4]   Conerly (2009).

straightforward. Finally, there do not seem to be significant barriers to building structures on the Moon in which atmospheric pressure and temperature could be maintained for human life, although they might have to be underground, for a number of reasons; but humanity is ingenious and adaptable, as we have seen throughout this book, and despite these engineering challenges, space architects have proposed inflatable structures, tunneling habitat-construction machines, the use of solar furnaces to bake construction bricks from lunar soil, and even crater-sized habitat designs that might work to colonize the Moon.[5] While the Moon lacks many resources, it does appear to contain significant amounts of helium-3, or tralphium, that may serve one day to fuel advanced fusion engines for spacecraft.[6]

However, there are good reasons to go farther out than the Moon in our first colonization of space. In his popular 1997 book, *The Case for Mars*, Robert Zubrin highlights a number of disadvantages in colonizing the Moon, including the fact that all nitrogen for agriculture would have to be brought from Earth or asteroids at tremendous cost, the complete lack of atmosphere, and the rather alien 28-day diurnal cycle in which any building materials would be exposed to repeated cycles of 28 Earth days of daylight (over 93°C/200F) followed by 28 days of night (less than -128°C/-200°F), which would cause severe engineering challenges. As we will see in the next section, Mars offers many more natural resources. But most important is that Mars is farther away; the Moon, Zubrin points out, is too close. A dependence on the Earth – politically, psychologically, materially and culturally – would probably constrain the full human engagement with adaptation to a new habitat on the Moon, which is only four days away by conventional spacecraft. In contrast, it takes months to reach Mars with conventional spacecraft, and a return trip would be a year's commitment. To really take up the extraterrestrial adaptation (to use our term) humanity should be bolder, and go directly to Mars.

Where Can We Go? Mars

Though it is farther away than the Moon, Mars is far more inviting, rich with natural resources and suited to human colonization. This world can provide the resources for many future generations, and technologically it is within our grasp.

Mars has held human attention and imagination for some time; the Red Planet was known to Ancient Egyptian, Babylonian, Greek, Chinese and Indian astronomers. In the 17th century Johannes Kepler (1571–1630), working with some measurements from his mentor Tycho Brahe (1546–1601), estimated the distance from the Earth to Mars,[6] and in 1610 Galileo Galilei became the first person to observe Mars through a telescope. By the late 1870s, Giovanni

---

[5]   Cohen (2002).
[6]   Duke (2006).

Schiaparelli (1835–1910) mapped a series of channels on the planet's surface as he observed them through a telescope, naming them after major river systems on Earth.[7] During the late 19th century, these *canali*, or channels, captured the attention and imagination of the American astronomer Percival Lowell (1855–1916), who dedicated much time and expense to the observation and mapping of the planet's surface as seen through his telescope. Lowell's observations stimulated much of the early science fiction regarding Mars, as he wrote that Schiaparelli's 'canali' had been excavated by intelligent beings attempting to save their dying, arid world by diverting melt water from the poles down to desiccated regions. Talk of a possible race of Martians stoked so much interest and speculation among turn-of-the-century astronomers that some began searching for other signs – such as the presence of gaslights – that Mars might be inhabited by intelligent beings.

It took nearly a century to learn the truth; when the Mariner 9 spacecraft orbited Mars during the winter of 1971–72, it revealed that while there were channels on the surface, they were dry river channels, not artificial constructs.

What else do we know of Mars? What makes it so compelling? First, the planet has a history much like our own. Second, it has resources that can fill our needs for water, breathing gas, food, and the regulation of temperature and pressure.

While today Mars is rather cold and airless, it is much warmer than the Moon, and long ago it was even warmer and much more like Earth. The early geological history of Mars is much like that of the Earth, having originated its own Achaean Eon after it formed through an accretionary process.[8] Like Earth, Mars also experienced heavy bombardment from meteorites and asteroids during the first two billion years of its life, and as on Earth, these may have contributed to melting and differentiation of the crust. And, like Earth, Mars had an early dense atmosphere, probably including water vapor, resulting from both volcanic activity and the delivery of other gasses to the surface by asteroid and meteor impacts. Through this time, Mars' primitive atmosphere most likely produced a greenhouse effect, trapping heat and warming the planet. The canyons and 'canali', then, date to somewhere around this time, when water flowed on the surface of Mars, something we can be sure of after nearly four decades of intense investigations.

The evidence is found many ways, including the comparison of *fluvial* (water-driven) terrain features on Earth with similar features on Mars. Sometimes, evidence has been unexpected, as in 2006, when NASA's Spirit rover developed a mechanical problem that forced the rover to drag one broken wheel through the ground like a plow, revealing mineral salts and silica deposits, both of which are normally formed only in watery environments. Additionally, the Mars Reconnaissance Orbiter detected several subterranean deposits of water ice using an imaging technique involving ground-penetrating radar, and more recently

---

[7] Taton et al. (2003): 109.
[8] Levin (2010): 22.

the Phoenix Lander found both calcium carbonate and perchlorates in the soil, again strong indicators of liquid water in Mars' past. The perchlorate is particularly interesting as it is a source of nutrients for some extremophile microbes on Earth, and may serve to trap water particles from the atmosphere in the soil. So, there is plenty of good reason that water once flowed on Mars, and we know where much of that water is today; in the north polar ice cap, which, according to one study, contains more water than the American great lakes combined.[9,10]

But whereas Earth retained its flowing water and in time developed an atmosphere that supported land-based life, on Mars something else happened. On Mars, not long after its own late heavy bombardment, about four billion years ago, the incipient atmosphere appears to have dissipated into space, leading to rapid evaporation of the surface waters and the freezing of the rest.[11,12] Why the atmosphere dissipated is currently unknown, but it could have happened if Mars had been grazed by an asteroid or other space debris massive enough to 'suck away' the atmosphere, or struck directly by something large enough to boil off the atmosphere.[13] In either case, as atmospheric pressure plunged, liquid water would have boiled away into free hydrogen and oxygen molecules, and some of the oxygen would have chemically bound with the iron present in the soil, explaining the iron oxides that give Mars its salmon- and rust colors. With a much thinner atmosphere unable to retain the warmth generated by solar irradiation of the planet's surface, temperatures would have dropped to those of present day Mars; lows to about –80°C (–110°F) and highs to about +27°C (+81°F).

For future human colonists of Mars, this planetary history is important. Of all Mars resources, the abundance of water is perhaps most important. Humans, we know, need water, breathing gas, food and temperature/pressure regulation. As on Earth's Moon, Mars water can easily be processed to both drinking water, and

---

[9]   See http://www.universetoday.com/23932/lots-of-pure-water-ice-at-mars-north-pole/.

[10] The Martian south polar ice cap also has some water ice, but is largely dry ice; see Bibring (2011).

[11] Often abbreviated in astronomical literature as LHB, this refers to a period in the development of our Solar System when there appeared to be an intense period of asteroid and comet bombardment of the inner solar system. From petrological studies of our Moon, it appears that this event occurred approximately 700 million years after the planets formed, or about 3.8 to 4.2 billion years ago. There are various theories for the cause of this event, the latest of which seem to be focused on the gas giants of the outer Solar System exerting gravitational influence on the various asteroids and other material in the Kuiper Belt, and sending them hurtling towards the Sun, thereby into the orbital paths of planets of the inner Solar System. The resulting effect was a period of heavy asteroid impact and cratering of the inner planets.

[12] Beech (2009): 134.

[13] Zubrin (1997): 120.

water for agriculture – and that agriculture would be easier on Mars than the Moon because Mars is richer than the Moon in fixed nitrogen. And while the atmosphere of Mars is thin, and of nearly 100% unbreathable carbon dioxide (rather than breathable and about 20% oxygen on Earth), that carbon dioxide itself can easily be processed into a wide variety of products, including breathing oxygen. As mentioned, Robert Zubrin has pointed out that Mars contains plenty of resources for human settlement, including soil that can be transformed into building materials, and, in addition to this, it is far enough from home to stimulate real, new human populations, "a new branch of human civilization".[13] Zubrin has also convincingly argued that the farther we go at first, the easier it will be to learn about going yet farther; the farther we go, then, the farther we will *be able* to go.

There are people today actively planning to go: members of the Mars Society itself (established by Zubrin) as well as many other, smaller groups including the Mars Homestead Project, based in Boston, whose goal is to "design, fund, build and operate the first permanent settlement on Mars".[14]

Whoever first moves to Mars, they will find a planet different from Earth in some ways, but similar in others. The Martian day, for example, is 1.03 Earth days, and Mars' axial tilt is much like that of the Earth, giving the planet seasons. The year is longer, however, at 686.6 Earth days, making each seasons on Mars about twice as long as long as on Earth.[15] Plants would grow well in greenhouses that simply concentrated and pressurized the Martian atmosphere, and they would provide food, organic debris, and oxygen. Mars greenhouses would naturally filter out much of the dangerous ionizing radiation that presently reaches the Martian surface, and they could be pressurized and heated so that people could work in them in shirtsleeves. It is clear that with water and breathing gas requirements met, human food requirements would also be met. And there are no barriers to the construction of homes, entire farmsteads, and, eventually communities; Mars is rich in the raw materials needed to make basic materials such as metals, glass and a kind of ceramic.

In the long term, some have more ambitious plans; terraforming Mars has been proposed by many to make it even more conducive to human life. Terraforming would essentially begin with attempting to thickening the Martian atmosphere, which would warm the planet, causing carbon dioxide and water vapor to outgas from the soil, further warming the planet and thickening the atmosphere. While current estimates suggest that this would take centuries, after that time, when sufficient atmosphere exists to better protect against the sun's ionizing radiation, crops, either adapted or engineered to Martian soils, could be grown in the open, without the need for pressurized green houses. Once plant life can survive and thrive on the surface, free oxygen will begin to accumulate in the atmosphere, just as happened in the ancient days of the Earth. Ultimately,

---

[14]   See www.marshome.org.
[15]   Barlow (2008): 21.

once oxygen levels reach levels of 18–21%, the atmosphere on Mars could be breathable. The natural timeline for this is on the order of thousands of years; however, there are ways in which the process could be accelerated artificially. If humanity were to carry out this project, it would surely be one of the more extended episodes of niche construction in the career of our species.

These are all interesting ideas, and enough of them are simple enough that they could work; humanity can colonize Mars. But as we have seen from human history and prehistory in chapters 3 and 4, our capacity to colonize Mars is not guaranteed to be preserved through time. We could lose our technical capacities in a dark age, of the same kind that has interrupted all the ancient civilizations. We should not panic, but we should make mature decisions, and they include ensuring a future for humanity by continuing what humanity has done in the past to ensure a future: expansion and colonization of new habitats, by the well-known processes of evolutionary adaptation.

Farther Afield: Worlds Beyond our Sun

In the long run, can we go farther? For a long time, the answers came purely from science fiction; it was only that lens that allowed us to imagine planets beyond our Solar System, in tales of faraway planets circling faraway stars. Many of those imaginary planets, of course, appear much like our Earth, though with an extra sun or moon or two in the sky to let the audience know that they are far from home. Although nobody had shown that there *could not* be planets around other stars, there simply was no evidence for them. Even with increasingly powerful light telescopes, the direct observation of a planet orbiting another star would been impossible because the light reflecting off of a planet would be impossible to see compared to the amount of light being emitted by the star itself. But in the past decade, we have added science to science fiction; today, with special methods, we can indeed 'see' real planets beyond our Solar System.

This is because of the work of astronomers Geoff Marcy and Bruce Campbell, who in the 1980s devised a technique to measure the gravitational force exerted by a planet on a star. Known as the Doppler method,[15] the technique relied on detecting a slight gravitational "wobbling" of the host star as a planet (or planets) orbited the star. The method is commonly used today; the *exoplanet.org* website, maintained by the California Planet Survey, currently lists 499 known expolanets (Figure 8.1 shows the masses and distances of these *exoplanets* known to science up to 2009). With the advent of the Hubble space telescope the search for "Super-Earths" began; rather than gas giant planets (like Jupiter), these are more Earthlike in that they are rocky and have a mass between 1–10 times that of the Earth (Jupiter, which is 318 times more massive than the Earth).[16] Some of

---

[16]   Folger (2011): 30–39.

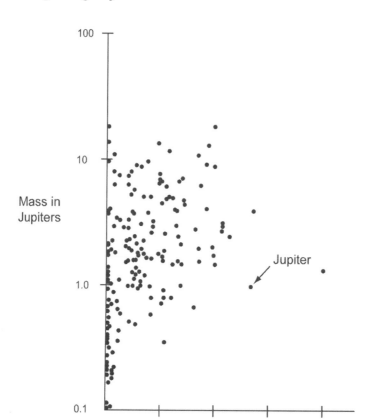

Distance from Exoplanet to its Sun in Astronomical Units

Figure 8.1. Exoplanets Known by 2009. Adapted from Figure 2.4 (p.28) of Eales (2009).

these worlds, it seems, must be habitable. And we have only just begun to look; in 2009 NASA began the Kepler mission, a systematic study developed to detect Earth-sized worlds in orbit around other nearby stars. Data from Kepler's photometry survey (different from the Doppler method) recently suggested that up to 1235 new Earth-sized planets have been discovered, orbiting some 997 host stars. In the terminology of astronomy the Habitable Zone, or HZ, also known as the "Goldilocks zone," traditionally referred to the distance from a star where liquid water could exist and surface temperatures were in a range where Earth life could survive. Seth Shostak, the contemporary astronomer most associated with the Search for Extra-Terrestrial Intelligence (SETI) initiative, believes that there could be "at least 30,000 habitable planets ..." within a thousand light years of

Earth.[17] It is important to mention here that these are just the planets that might be in a range to support life as we on Earth know it; we must not discount the possibilities that life may have arisen on planets or even asteroids that we would have never considered possible, in forms we cannot even contemplate.

In any case, today we are in the astounding position of being able not only to imagine other, distant planets and stars, but also to study them directly; we can know their mass, distance from sun, temperatures and other critically important variables. In evolutionary terms, the question is simple: could the planet supply our needs for water, breathing gas, and the regulation of temperature and pressure? If so, or if we can manipulate the environment to fulfill these needs, we can start to make a shortlist of planets to visit when the technologies are available; our shortlist might be so exciting, of course, that it actively stimulates the development of the technologies needed to visit the planets of other stars.

What Will Change in Human Extraterrestrial Cultures?

Wherever humanity moves, it is easy to say that Earth and off-Earth cultures would diverge over time, eventually 'splitting' to become new cultures as geographically separated life forms diverge genetically, ultimately becoming new species. But can we specify what would change between Earth and off-Earth cultures, whether the off-Earthers are on Mars, or some other distant body? That is difficult to answer, but a good place to begin, or at least to stimulate thinking, is the list of 'cultural universals' published by cross-cultural anthropology pioneer George P. Murdock in 1945, a list of "items, arranged in alphabetical order to emphasize their variety, which occur, so far as the author's knowledge goes, in every culture known to history of ethnography."[18] While in academia each of these items could be challenged and defined *ad infinitum,* and we do not propose them as concrete units; rather, they give a flavor of the variety of elements present in all cultures, and it is an illuminating exercise to contemplate how these (and other) specifics would diverge over time, among Earth-based and off-Earth cultures.

age-grading * athletic sports * bodily adornment * calendar * cleanliness training * community organization * cooperative labor * cosmology * courtship * dancing * decorative art * divination * division of labor * dream interpretation * education * eschatology * ethics * ethnobotany * etiquette * faith healing * family * feasting * firemaking * folklore * food taboos * funeral rites * games * gestures * gift-giving * government * greetings * hairstyles * hospitality * housing * hygiene * incest taboos * inheritance rules * joking * kin groups * kinship nomenclature * language * law * luck superstitions * magic * marriage * mealtimes * medicine * modesty concerning natural functions * mourning * music * mythology * numerals * obstetrics * penal sanctions * personal

---

[17] Shostak (2011).
[18] Murdock (1945).

names * population policy * postnatal care * pregnancy usages * property rights * propitiation of supernatural beings * puberty customs * religious ritual * residence rules * sexual restrictions * soul concepts * status differentiation [different rights and responsibilities according to gender, age, and so on] * surgery * tool making * trade * visiting * weaning * weather control

Remember, it is the specific values that vary among cultures in these (and other) cultural 'items', such that while both Japanese and Canadian cultures have certain ideas about the treatment of visitors, for example, both have certain ideas regarding visitors, visiting, hosting, and so on. In terms of human adaptation to space and its longer-term effects, it is clear that while there will be slow biological change between Earth and non-Earth populations, we can expect that, just as on Earth in the past millennia, it will be largely the divergence of specific values of these (and other) cultural 'items' that will result in Earth Culture and Space Culture diverging over time. To accommodate new values, phenomena, and traditions, new words will accumulate (both on Earth and off of Earth), further separating these cultures. That is natural; humanity, as we have seen repeatedly, diverges according to habitat. How much divergence will be tolerated by one of the other, and what their reactions to these divergences will be, are generally unpredictable, but in their specifics might be of interest. For example, how might musical tastes diverge? They certainly change quickly on Earth, leading to a wonderful range of musical traditions across the Earth and through time.

One example of how the contents of cultural elements, as indicated above, can diverge, is seen in the case of naval officers and land-based scientists aboard naval sailing vessels of 19th century America. On voyages of scientific discovery, historian Helen M. Rozwadowski has written, the scientists kept land time even when aboard ships, and the officers, of course, kept ship's time ('watch time'), even when ashore.[19] It is the accumulation of small differences, like this, that ultimately result in cultures so different from one another that – despite a great deal of anthropological hand-wringing about the difficulties of drawing a line around a 'culture' – human cultures do indeed differentiate. And, with this case in mind, it is easy to imagine how values and customs would diverge over time, 'attached', one might say, to such a fundamental issue as different conceptions of time. Another example of small differences that might well arise between Earth and off-Earth cultures is in the field of folk beliefs, or superstitions. However careful our plans, for some time the colonial life will be more dangerous than Earth life, and on Earth, at least, anthropologists have observed that in environments perceived to be dangerous (and/or actually dangerous), people are often significantly more superstitious than in environments either less dangerous or perceived to be less dangerous. While there is no telling how precisely such superstitions and taboos will 'shake

---

[19] Rozwadowski (2005): 208–209.

out' – who knows what Martian miners will consider taboo – the point is that we can expect superstitions to arise in such relatively dangerous environments as extraterrestrial colonies, and they will be one of the multitude of small cultural elements that ultimately drive a wedge between off-Earth and on-Earth cultures. For example, below we list a few superstitions held by traditional New England fishermen in the 1960s:[20]

- Don't turn any hatch cover upside down – causes bad luck.
- Don't whistle – causes windy weather.
- Don't serve beef stew aboard – causes stormy weather.
- Calm weather indicates approaching poor weather; remain apprehensive.
- Always refer to the boat as "she".
- Don't set out to sea after seeing a rat ashore.

It is too early to speculate on how such cultural differentiation will actually play out, but that is not the lesson. The lesson is that we should expect such change to occur, and we should recognize it as adaptation to new circumstances when it does occur. Children move away from home, and off-Earth people will eventually not return to Earth, at least not all of them, and they will become different. That, it seems, considering that the Universe is characterized by change, is natural.

What Will Change in Human Extraterrestrial Biology?

It is often waggishly suggested that as humanity increasingly automates factories and other production facilities, our species will become a large brain with a single finger for pushing buttons. In the near term this is very unlikely as there is a long evolutionary heritage to our genome that simply precludes such large, organizational changes except over very long periods. Still, if just one gene or variant of a gene spreads significantly in a population, that is evolution, and we can take a moment to think about this issue. Can we imagine what kind of biological characteristics will change in time, in extraterrestrial humans? Not really. As with culture, discussed above, human biological evolution cannot be precisely predicted. For example, it was recently discovered that lactose tolerance – the ability to digest milk into adulthood, which is absent in many people worldwide but common in Europeans as well as cattle-herding people of Africa – is a relatively recent genetic variation, an instance of evolution in humanity's digestive system.[21] This significant change – which results in different diets, food preferences, and other far-ranging aspects of language and culture that in part differentiate East Asian, European and some African cultures – would have been very difficult to predict, say, ten thousand years ago, before the advent of

---

[20] Poggie and Gersuny (1976): 355–356.
[21] Tishkoff et al. (2007).

pastoralism and other forms of domestication that favored the consumption of cow milk.

Another factor that makes human biological evolution difficult to predict is that a great deal of the natural selection that has until very recently shaped human biological evolution has been effectively 'buffered out' by inventions and technology, such that genes that in generations past would have been eliminated by natural selection, persist today. How this will affect future generations of space colonists is unknown, but likely to be a significant field of research. It should be remembered that while we might think that screening out all such deleterious genes from space colonizing populations would lead to a hyper-pure, 'clean breed' of humanity, such a conception has long been shown to be a fool's paradise. Genetic sterility, it turns out, is genetic suicide, since if all members of a population are genetically identical, or near so, they will all, ultimately, be the victims of a single genetic 'sweep', when – for instance – an unknown and unpredictable virus finally *does* arrive, wiping out the entire population. This is not science fiction, but fact, and today ecologists monitor the health of a given species not by its genetic uniformity, but by its genetic diversity. Much concern has been voiced, also, about the possibly catastrophic effects of producing low-variation, genetically very similar crops and even animals (such as fish) that could all be wiped out in single infections of unknown organisms.

These examples should cause us to move away from the question of just how human biology will change over time and to increasingly examine how we will adapt to new habitats in ways that will maintain the genetic health not only of the human species, but of the many domesticated plants and animals that we will take away from Earth to be our food stock. And we will have to be very aware of the many normally-invisible species with which we humans (and our domesticates) coexist, as Alfred W. Crosby has pointed out, writing that, after the physical environments of space, "The next most dangerous enemy in the immediate foreseeable future will be humans themselves and the irrepressible organisms they will carry with them."[22] It is clear that the field of physical anthropology, which has the best understanding of human demography and population genetics, will be critical to designing an adaptive approach to biological evolution in space.

Some essential issues in human biological evolution off of Earth can be sketched out, but we are only just beginning to study them.

Mutagenesis

Not only does space radiation 'zap' our genes (and those of our domesticated plants and animals), thereby creating potentially dangerous mutations; in some cases, new variations come from (a) 'mutator genes' already riding the DNA, (b)

---

[22] Crosby (1985): 210.

the failure of DNA-repair mechanisms, and (c) other environmental influences.[23] All of these will have to be considered when exposing human DNA – and that of all of the domesticated plants and animals that go with us into space – to new environments, whether on the Moon, Mars, in orbital colonies, and so on.

Founder Effect

Founding populations of colonies are typically small, and their genetic constitutions will have significant effects in the future of the colonial population.[24] This is a major issue that leads to nightmarish visions of discriminatory colonist-candidate screening – as in the dystopian film *Gattaca* – but in reality the concept of a 'super race' is genetically suicidal, as mentioned earlier when we discussed the health of populations as being measured by diversity rather than uniformity. To best understand and plan for founder effect issues, human population geneticists (who work with theoretical models) should collaborate with cultural anthropologists (who know much about the real world of cultural behavior and variation from patterns of behavior).

Holobiont Evolution

Humans – and most other organisms – coevolve with a multitude of microbial species, which will accompany people (and humanity's domesticated plants and animals) into space. The *holobiont* (also known as the hologenome[25]) refers to an organism and all of its microbial, coevolving companions in evolution. Since many of these microorganisms are important to such processes as digestion, stimulation of the immune system, vitamin processing, fiber breakdown, and others,[26] it will be necessary to make a thorough cataloging of human microbial populations so that we will know who our 'microcastaways' will be, and how they might affect future evolution.

Communicable Disease

Colonies are characterized by small populations living closely together, ideal conditions for the spread of communicable disease, and it will be quite a while before off-Earth populations will be resilient enough to sustain catastrophic disease (naturally, efforts will be made to prevent this in the first place). As a long-term concern, disease is far more worrisome than the space aliens, meteorites, or even sabotage we are expected to worry about in Hollywood's

---

[23] Friedberg (2006).
[24] See Rosenberg et al. (2005) for a paper showing that even how to model such issues only recently been addressed.
[25] Rosenberg, Sharon and Ziller-Rozenberg (2009).
[26] Rosenberg and Ziller-Rozenberg (2011): 60–63.

version of the human future in space. This issue can be approached with a review of the history of communicable disease in smaller-than-modern populations,[27] and design of all manner of preventative, quarantine and treatment plans.

Minimum Viable Population

Every species has a minimum number of interbreeding individuals that must be maintained; if populations fall below that number (the MVP or 'minimum viable population'), the ill effects of inbreeding – including many genetic disorders – will begin to cause problems.[28] Space colonists will have to maintain the MVP by a number of methods, including new social arrangements for finding and cohabiting with marriage partners. Such social arrangements will affect the design of housing, transportation and other elements of off-Earth settlements.

Domesticate Health

Humanity will, as mentioned, not be alone on the adaptation to space; there will be our microcastaways as well. And, just like Polynesian voyagers of the ancient world, we will take with us a wide range of plant and animal domesticates as staple foods, condiments, and for other reasons. All of *their* population, biological and genetic issues, as well as those of their own holobionts, will have to be considered as well. For this reason, space colony planners should work closely with both farmers who know such plants and animals, and microbiologists who know their microbial cohort.

Finally, we should be prepared for the strong possibility that human biology will change to some degree in ways that we cannot yet imagine, because we do not know the environments we wish to colonize entirely. Furthermore, we should be prepared for the possibility that, despite our best efforts, natural selection will once again take a toll on human lives, until adaptations are invented to cope with new selective pressures. As mentioned before, however, our ancestors lived with such risks, and we accept thousands of deaths yearly from mere car crashes as simply 'risks of life'. If we are going to assume such risk, why not assume it in the pursuit of the extraordinarily important goal of spreading humanity beyond the cradle?

In Chapter 3 we reviewed Terrestrial, Technical and Cognitive adaptations. In Table 8.1 we include what we term the 'Extraterrestrial Adaptation'. (In the table, – means the characteristic is decreasingly significant, + means the characteristic is increasingly significant.)

---

[27]   For example, archaeologists have studied the demography and other variables of disease in ancient cultures who – even in early urban times – lived in smaller populations and lower population densities than in modern civilization: see Roberts and Manchester (2005).
[28]   Thomas (1990).

Table 8.1. Placing the Extraterrestrial Adaptation in a Larger Context.

Adaptation	Biological Change Involved: Structural/ Anatomical	Biological Change Involved: Metabolic/Process	Cultural Change Involved
Preadaptation	– canine size	– selective pressure for robust body	+ social complexity
Pair-Bonding: by 7mya	– sexual dimorphism	+ selective pressure for complex social communications	+ social complexity
Terrestrial Adaptation			
Bipedalism: by 7mya	+ bipedal anatomy – arboreal anatomy	+ thermoregulatory efficiency	+ territorial range + adaptive econiche plasticity
Technological Adaptation			
Tool Use: by 2.5mya	+ finger–thumb opposability	+ hand/eye coordination	+ enculturation time
	+ brain volume	+ caloric needs	+ econiche breadth
	+ body stature	+ caloric needs	+ econiche breadth
	– tooth size	– digestion requirements (food pre-processed by stone and, later, fire)	+ econiche breadth = reliance on technology = reliance on culture to teach complex tool use = decoupling of behavior from anatomy
Cognitive Adaptation			
Modern behavior: after 100,000BP in *Homo sapiens*	+ brain volume	+ caloric needs	+ econiche breadth & or + econiche specificity & or active Niche Construction
	+ brain volume & brain architecture complexity	= Lowered Neuronal Activation Threshold	= + cognitive variation/ complexity = + behavioral variation
	– tooth size	= more reliance on technology to process foods = more complex social interactions	= + cultural complexity and enculturation time
	– body robusticity		= + cultural complexity and enculturation time

Table 8.1 cont.

Adaptation	Biological Change Involved: Structural/ Anatomical	Biological Change Involved: Metabolic/Process	Cultural Change Involved
Extraterrestrial Adaptation			
Beginning at present	early: none later: largely unforeseeable, but depends on selective environment, e.g. Mars surface-vs-Mars orbit + biological differentiation between Earth and off-Earth populations	unknown and under investigation	+ data management + evolution manipulation + spatial sense + temporal sense + cultural differentiation = yet greater reliance on technology but also adaptive qualities of culture and intelligence

The "Other"?

Traditional socio-cultural anthropologists spend their careers, and often their lives, in pursuit of an intimate understanding of the people of their studies. In self-reflective anthropological discourse a given people, traditionally from a non-Western society, who are the subject of an ethnographers study are generally referred to as "the other." Anthropological fieldworkers have written volumes about the lives and beliefs and social structures of the other, ethnographers have striven to understand the various "other" peoples and cultures and to illustrate the basic commonalities between our western society and that of a people whom an anthropologist has studied.

Almost 2500 years ago the Greek philosopher Metrodorus posed the question, "is it reasonable to suppose that in a large field that only one shaft of wheat should grow, and that in an infinite Universe, to have only one living world?"[29] We know that occurrences are rarely unique in nature, and that nature abhors a vacuum. Given our understanding of life and how profligate and diverse it is on Earth and given recent discoveries in astronomy that provide for tantalizing possibilities, it is not at all unreasonable to allow for the possibility that life has arisen elsewhere in our galaxy. Is there a cosmic "other" then? Is there an extraterrestrial life form that exhibits intelligence? An extraterrestrial life form that has developed a civilization? If so, where are they? How do they look and act? Will we ever meet them, and if we do, can we ever hope to communicate with and understand them?

---

[29] Zubrin (1999): 247.

When positing on the nature of advanced extraterrestrial life and a spacefaring extraterrestrial society, some authors go to great lengths to discuss their behavior and how they might be inclined towards us, the lifespan of an advanced technological civilization, and so on. We make few suppositions of this nature below. To begin, at this stage in our understanding, all bets are off. If we were to encounter an advanced spacefaring species, we would be confronted with an intelligence and a philosophy that we have never before encountered, and one that may not even be possible for us to understand. We cannot assume that an alien species would be motivated by what we are, nor would share any universal system of values with us, or perhaps even recognize us. Carl Sagan wrote that extraterrestrial intelligence would be "elegant, complex, internally consistent, and utterly alien."[30] At the moment, our understanding of life is confined to the forms, plentiful and varied though they be, found only on our home planet. Assuming life has arisen elsewhere in our cosmos, it is almost certain to be very different from anything we understand.

Biologists define life by four general processes: growth, reproduction, responsiveness, and metabolism. Scientists are in general agreement that if a collection of organic molecules increases in size, if it makes copies of itself, if it somehow responds to its environment, and if it somehow incorporates elements from outside its structure and converts them in a series of controlled internal chemical reactions to compounds needed to grow, reproduce or physically respond to changes in its environment, it is alive.[31]

In the early 1960s Frank Drake conducted the first search for radio wave signals from potential extraterrestrial civilizations at the National Radio Astronomy Observatory in Green Bank, West Virginia. This began the international astronomical effort known today as the Search for Extra-Terrestrial Intelligence, or SETI.[32] In 1961 when the he was asked to chair a meeting on the

---

[30] Sagan (1980): 296.

[31] Bauman (2009).

[32] The search for extraterrestrial intelligence began as Project Ozma (after the "Wizard of Oz") in 1960 with Professor Drake searching for radio signals from stars about 11 light years from Earth. Since that time, SETI has come to encompass a multitude of efforts by various groups to search for signals that may have been produced by an alien technological civilization. SETI efforts have been funded periodically by NASA, and have utilized the Arecibo observatory in Puerto Rico, the National Radio astronomy Observatory in Green Bank, West Virginia as well as Ohio State University's former "Big Ear" radio telescope, which conducted the longest-running SETI program in history, from 1973–1995. Funding continues to be a difficulty for SETI studies, and as a way of defraying costs, and analyzing many more data sets that would otherwise be possible, the University of California at Berkeley operates SETI at home (SETI@home), a system whereby anyone who has a home computer connected to the internet can assist astronomers in the SETI effort. The system uses large numbers of internet-connected computers to assist in the data analysis phase of SETI. The free software may be downloaded from http://setiathome.berkeley.edu, and the program downloads and analyzes radio telescope data.

detection of extraterrestrial intelligence by the National Academy of Sciences, Drake developed his famous equation designed to estimate the number of advanced technical civilizations in our galaxy.[33] In 1980, Carl Sagan popularized this equation in his television series *Cosmos* to point out to television viewers across the world that our own galaxy might very well be teeming with not only life, but likely other advanced technological civilizations. Professor Drake persuaded astronomers and other interested researchers to think seriously about the possibility of other intelligent life in our galaxy, and Sagan persuaded the common man to think about that same possibility, and to consider the implications for our own existence, thereby raising our consciousness.

The Drake equation has many variations and through the years different researchers have estimated different values, and so have come up with many different results. The original equation and all of its spin-offs have some intrinsic problems, as we will see, but the original Drake equation is

$$N = R_* \, f_p \, n_e \, f_l \, f_i \, f_c \, L$$

where N represents the number of technological civilizations in the Milky Way galaxy, R_* *is the rate of new star formation in our galaxy*, f_p is the fraction of stars that have planetary systems, n_e is the number of planets in a given system that are capable of supporting life, f_l is the fraction of such suitable planets where life actually arises, f_i is the fraction of such planets where intelligent life develops, f_c represents the fraction of the previous term's planets that develop a civilization capable of broadcasting its existence into space through radio waves as we on Earth do, and L represents the lifespan of such an advanced civilization.

In 1961 Drake used the following values for the terms:

R_* = 10 per year
f_p = 0.5 (assuming half of all stars in the galaxy have planets)
n_e = 2 (stars with solar systems will have two life-friendly planets)
f_l = 1 (100% of such planets will develop life)
f_i = 0.01 (1% of which will be intelligent life)
f_c = 0.01 (1% of which will be able to communicate)
L = 10,000 years (the lifespan of such an advanced civilization will last 10,000 Earth years).

Using these values Drake and his colleagues values found that $N = 10 \times 0.5 \times 2 \times 1 \times 0.01 \times 0.01 \times 10,000 = 10$. With values that could only be guessed at in 1961, Drake reasoned that there were about ten civilizations in our galaxy that use radio waves to communicate, which we should be able to detect using radio telescopes. In 1999 Robert Zubrin used different values, including a value of f_l of some 50,000 Earth years, and came up with 400 such civilizations in our galaxy.

---

[33] Sagan (1980): 299.

Sagan's version was slightly different:

$$N = N_* \, f_p \, n_e \, f_i \, f_i \, f_c \, f_L$$

the terms being much the same, with the exception of replacing R_* (the rate of star formation) simply with N_* (the current number of stars in the Milky Way galaxy) and with the addition of f_L. In his presentation of the data, Sagan attempted to show how the terms could be variable, and focused especially on the value of f_L, the potential lifespan of a civilization. If alien civilizations, like past civilizations here on Earth are prone to collapsing after a certain amount of time, or if they trend towards self-destruction, as many felt that we were on the brink of in the 1980s due to the build-up of nuclear arsenals by the United States and the Soviet Union, f_L would be very small. If that were the case Sagan observed that there "might be no one to talk with but ourselves, and that we do but poorly."[34]

The real problem with the terms of the equations, as many have observed, is that we can really only guess at the values for most of them. It is encouraging however that we are further along in our understanding of the Universe that we were in 1961, and now, for example thanks to NASA's Kepler, we have a much better idea of the frequency of planetary systems around other stars, f_p and even some idea of their possible composition and perhaps ecology, making n_e less of an unknown. For the rest of the terms we really have absolutely no idea of their numbers, and many of our estimates are hopelessly anthropocentric, or biased towards our own experience and understanding. If a civilization spread beyond its home planet and throughout space to many other worlds, there need not be a finite lifespan imposed on the civilization at all, which is what makes moving beyond our own world so tantalizing, so hopeful. There is currently no data from which to derive reasonable values for f_L although that will most likely change one way or another once we begin a comprehensive program of exploration and study of our own Solar System.

The values of f_i and f_c are completely unknown, possibly unknowable without far more data, and may themselves not even be valid terms. We are assuming far too much to think that we would even be able to recognize an alien intelligence, a civilization or its artifacts. Without a better understanding of how and where life can arise, what other forms intelligence can take, and what form an alien civilization might take, any number arrived at for N is next to meaningless. But what the equation has really done, what Professor Drake intended it to do was, even in the absence of data, to help us define and think more about the various factors necessary to predict the likelihood of extraterrestrial civilizations in our galaxy and to stimulate thought and debate on the subject.

We have even attempted to communicate with any "others" who might be out there and might, in turn, be interested in us.

---

[34] Sagan (1980): 301.

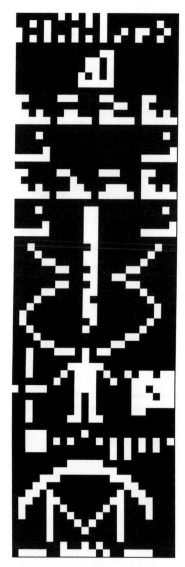

Figure 8.2. Binary Image Transmitted to Deep Space by the Arecibo Antenna. NASA public domain image.

In 1974, astronomers transmitted an image (Figure 8.2) as a 1679-bit binary signal from the Arecibo observatory in Puerto Rico toward the great cluster in the constellation of Hercules, Messier 13. November 16th, 1974, then, marked the first time we have deliberately sent a signal geared towards another intelligent species into space. In 25,000 years, when the signal final reaches its destination, the stars in Hercules will most likely be in a different position, but 25,000 years is a long time, and something or someone may receive the signal at some point. The message was designed by none other than Frank Drake and Carl Sagan.

The 1679-bit signal, Drake and Sagan hoped, would be recognized as the product of 73×23, both of which are prime numbers. If the signal were to be arranged in a 73×23 array, the picture would be seen. Arranged in a 23×73 format it would appear as a random collection of points, hopefully appearing meaningless. From there, Drake and Sagan hoped the recipients would be able to decipher the message, which describes the major elements of Earth, the DNA building blocks of humans, as well as which planet of our solar system we inhabited, and finally, a schematic image of the radio telescope itself. Donald Campbell, who was a research assistant at Cornell University, which runs the observatory, and worked at the observatory when the message was sent stated on the 25th anniversary of the transmission, has stated that that it was not a serious attempt to communicate with an extraterrestrial intelligence, but rather only a demonstration of technology.[35] No doubt Drake and Sagan felt, at least somewhat, differently.

In contrast to the Arecibo message, the *Pioneer* and *Voyager 1* and *2* spacecraft carried sophisticated messages (created by state of the art technology at the time) designed not as an attempt to contact an alien civilization, but almost as a

---

[35] Steele (1999).

"maker's mark" should something discover these probes at some point in both distant space and time.

It is arguable that any organism possessing spacefaring technology, as we understand it, would have had to develop a sophisticated understanding of physics, and should then be well able to understand mathematical concepts, and recognize that there is a basic order to the physics of the Universe, and therefore recognize that order in our symbology. But, of course, symbols such as letters or numerals are normally arbitrary, bearing little resemblance to what they signify, so it is difficult to say whether an alien civilization could make heads or tails of our messages, or vice-versa. Still, Drake and Sagan were optimistically looking to the commonalities that we would share with another species. They knew the differences would be vast, but thought it better to begin with the traits that we likely share, such as a similar chemistry, one that certainly would include hydrogen, one of the most common materials in the known Universe. One's reaction to the *Voyager* messages has much to do with one's capacity for optimism and imagination.

For all our cryptographic abilities, however, we again assume much. We are assuming that any intelligent recipients generally think the way we do, that they organize information in more or less the same manner that we do, that they would even be visual creatures. We are limited by our lack of knowledge of how an intelligence that developed on another world might "think" but we need not be limited by our imaginations. In our considerations of the nature of extraterrestrials we must be careful not to assume that life based on a completely different biology–with an intelligence arising independently in the Universe, equipped with a "technology" (to use the word loosely) potentially hundreds of millions of years more advanced than ours– would have anything but perhaps the most basic chemical elements in common with us.

As long as we are contemplating extraterrestrial life, one of the more exciting challenges for us is to imagine what such life might actually look like. If extraterrestrial life is built by DNA or some equivalent of it, we might hypothesize that life evolving on any planet would be governed by the need to overcome gravity, meaning that any sort of alien animal life would need a locomotor anatomy, something perhaps recognizable to us. Or perhaps it is not that simple. Astronomer Royal Martin Rees has posited that "there could be life and intelligence out there in forms we can't conceive."[36] We tend to think in terms of "animals" and "plants", perhaps occasionally even "fungi" when we think of the forms that life can take. And, the basis of all Earth life appears to be cellular, whether those cells are eukaryotes, prokaryotes or archaea, living things (we found in Chapter 2) that are either single-celled or multicellular organisms . But what if non-Earth life is built of something other than DNA, or even a close equivalent of DNA? An intelligence evolved from profoundly different biology would function very differently that our own.

---

[36] Leake (2010).

If we restrict our musings to intelligences recognizable to us, however, we could hypothesize that intelligent extraterrestrials species might have evolved as social and somewhat predatory life forms. If such a species were also aggressive and highly competitive, as Stephen Hawking and Robert Zubrin have suggested, we could easily be faced with hostile and intelligent alien versions of the worst aspects of our human selves,[37] and Hawking recently cautioned against actively broadcasting human existence to potential extraterrestrial civilizations, stating in the London Sunday *Times* that "we only have to look at ourselves to see how intelligent life might develop into something we wouldn't want to meet." Rather than advanced, benevolent extraterrestrials as depicted in much science fiction literature, Hawking continued that intelligent alien life might come to Earth "in massive ships, having used up all the resources from their home planet. Such advanced aliens would perhaps become nomads, looking to conquer and colonize whatever planets they can reach." This sounds familiar because the histories of the indigenous people of New Guinea, Australia, Africa and the Americas were all negatively – to put it lightly – affected by 'contact'. Humanity at large might be helpless in the face of such a contact.

As long as we are using humanity as our model for extraterrestrial intelligence however, there is another facet of our nature that needs to be considered: we humans have another side that is curious and inquisitive, has empathy for other forms of life, treasures companionship, and seeks understanding, enlightenment and growth.

This is our better human nature. In all of our musings of extraterrestrial intelligence, we may just as easily suppose that if an extraterrestrial civilization developed advanced technology and set out to the stars, they must have harnessed that better nature, progressing beyond their more basic instincts. Such a species would therefore not be bent on conquering the Earth, or stealing its resources. They might ignore us altogether, or they might attempt to contact us, perhaps even engage with us. For the moment, it is impossible to know.

Only a century ago, television, internet, space flight and modern computers were fantasies, almost unthinkable. What if there were an intelligent, technological extraterrestrial species that were a few thousand years more advanced in their technology than we are? What about a few million? Their technology might easily be incomprehensible to us; as Arthur C. Clarke commented, "any sufficiently advanced technology would be indistinguishable from magic".[38] And astrophysicist Paul Davies raises the issue that "even in more carefully crafted science fiction, alien artifacts appear recognizably as *machines*, in the 20th-century understanding of the term".[39] Any tools made by a non-terrestrial intelligence may not even be recognizable to us. Davies raises the possibility that such technology may not even be made of matter, but some form

---

[37]   Zubrin (1999): 215.
[38]   Gilster (2004): 163.
[39]   Davies (2010): 144.

of energy with no fixed size or shape, or, if such technology were perceivable to us, its purpose might be completely unapparent. Martin Rees has pointed out that some domains of reality must be well beyond the capacity of our own minds, and certainly our senses are limited, and the tools we use to examine the physical world in greater detail are devices that merely just extend those senses; electron microscopes, radio telescopes, stethoscopes, X-ray machines and so on. Our biology and our consciousness limit our ability to completely understand our Universe.

Despite the interesting possibilities raised by the Drake Equation, physicist Enrico Fermi's (1901–1954) question still nags: "Where are they? Where is everybody?" Could humanity be alone? If so, Martin Rees has pointed out, perhaps the most compelling reason to move outward beyond Earth is to preserve humanity and consciousness. Viewed in this light, the fate of humanity is profound. If we choose to move beyond Earth, first to Mars and then beyond, it is entirely possible that humanity and the plants and animals we bring with us will diversify into new varieties – cultures and species – with each world we move to. That would be natural, not unnatural.

Here There Be Dragons

Whether alone in the Universe, or one of many, we know that the future of humanity is not assured. Just as an infant matures away from the immediate world of here, now, and me, humanity can see beyond itself, its home planet, and make plans for a secure future. We are a proactive species that survives only because of conscious choices and active construction of places and ways to live. Today our choice is whether to remain in our infantile bubble or break free and move towards the edges of the maps. Written on the far corners of the charts of the explorers were often the Latin words *Terra Incognita*, representing unknown land, or "beyond here, there be dragons" representing the peril of these unknowns. Some pushed into those unknown regions, however, and today we are once again looking beyond the edge. But today we can see farther than any human before in our history has ever seen. As in the past, the unknown deters some, but others remain eager to explore: even without concrete plans to explore Mars by any space agency, people continue to invent and build, confident that we will eventually go (see Figure 8.3). It is time for us to explore again, and with great vigor.

The drive to explore still exists within each of us today. The restlessness of our prehistoric ancestors suggests that it is perhaps a fundamental aspect of our better human nature. It is that very nature that we must now decide to either harness, or remain preoccupied as most are today by our Earth-bound pursuits. We are on the cusp of a new age of heroic exploration, one rich with tremendous possibility, one that can be filled with fantastic discoveries and one that can lead toward the eventual colonization of space. If we set off on this path, those who choose to do so will almost certainly continue to grow, develop and thrive. As

Figure 8.3. Dr. Dava Newman Wearing the Biosuit. Image courtesy of Professor Dava Newman, MIT: Inventor, Science and Engineering, Guillermo Trotti, A.I.A., Trotti and Associates, Inc. (Cambridge, MA), Design, Dainese (Vicenza, Italy), Fabrication and Douglas Sonders, Photography.

our understanding continues to progress and our technology continues to build upon itself, we will become increasingly capable, we become able to accomplish what was once impossible, and moving farther and farther outward will become ever easier. If we choose this path, we will one day, perhaps sooner than we think, attain the stars. We will gain the ability to explore other solar systems where we can find new homes and new, unimaginable tomorrows.

The good news is that we can begin now, and everyone can contribute to the effort continuing the evolution of our species. We each have the ability to begin this great journey by taking small steps into the great cosmic ocean. We can encourage, join or otherwise support the many private efforts to develop spacefaring technology and all efforts to establish a permanent human presence on Mars, because Mars is the best place to begin building permanent, flourishing, off-Earth communities. We can change the way we think of space, just 60 miles above our heads, not just as the domain of the test pilot or engineer, but as a place for all, not simply a few. We can begin by thinking about how humanity has adapted in the past, and how our knowledge of adaptation can shape our future adaptation to space. But however we each choose to contribute to the larger goal of ensuring a future for humanity, we must move beyond the shores of Earth, and we must move before it is too late. The notion of the unknown still excites us, the mysteries that lie just beyond our small world still call to us. With a renewed and profound respect for life on Earth and perhaps elsewhere, we must again set forth; we must continue the journey. The successes and innovations of our past encourage us, the brightest future humanity has ever imagined beckons to us; like the mariners of ancient Polynesia we must once again look to the stars, and this time chart our course towards them. For our ancestors, our families and our species, it is time to set the sails again at last.

Bibliography

References to Chapter 1

Asimov, I. (1979). *A Choice of Catastrophes: the Disasters That Threaten our World.* New York: Fawcett Columbine.

Barnett, H.G. (1953). *Innovation: The Basis of Cultural Change.* New York: McGraw Hill.

Dobzhansky, T. (1971). Man and Natural Selection. pp. 4–18 in Bajema, C.J. (ed): *Natural Selection in Human Populations: The Measurement of Ongoing Genetic Evolution in Contemporary Societies.* New York: John Wiley & Sons.

Dunbar, R.I.M. (1998). Behavioural Adaptations. pp. 73–98 in Morphy, H. and G. Harrison (eds). 1998. *Human Adaptation.* Oxford: Oxford University Press.

Finney, B.R. and E.M. Jones (eds). (1985). *Interstellar Migration and the Human Experience.* Berkeley, CA: University of California Press.

Koonin, E.V. (2009a). The Origin at 150: is a New Evolutionary Synthesis in Sight? *Trends in Genetics* 25(11): 473–475.

Koonin, E.V. (2009b). Darwinian Evolution in the Light of Genomics. *Nucleic Acids Research* 37(4): 1011–1034.

Kottak, C. P. (2009). *Cultural Anthropology.* 13th ed. Boston: McGraw Hill.

McCurdy, H.E. (1997). *Space and the American Imagination.* Washington, DC: Smithsonian Institution Press.

Minkoff, E.C. (1983). *Evolutionary Biology.* Reading, MA: Addison-Wesley.

Pääbo, S. (2004). Ancient DNA. pp. 68–87 in Krude, T. (ed). *DNA: Changing Science and Society.* Cambridge: Cambridge University Press.

Raup, D.M. (1993). *Extinction: Bad Genes or Bad Luck?* New York: Norton.

Rees, M. (2004). *Our Final Hour: A Scientist's Warning.* New York: Basic Books.

Smith, C.M. (2006). Rise of the Modern Mind. *Scientific American MIND*, August 2006: 73–79.

Smith, C.M. and J. Ruppell. (2011). What Anthropologists Should Know About the New Evolutionary Synthesis. *Structure and Dynamics* 5(2). Online at http://escholarship.org/uc/item/18b9f0jb#page-1.

Smith, C.M. and C. Sullivan. (2006). *The Top Ten Myths About Evolution.* New York: Prometheus Books.

Tamm, D.J. (2006). *The Reinvigoration of the West through Outer Space Development.* Unpublished Master's Thesis, Center for European Studies, Jafiellonian University in Krakow.

Vermeij, G.J. (1978). *Biogeography and Adaptation.* Cambridge, MA: Harvard University Press.

Woese, C.R. (2004). A New Biology for a New Century. *Microbiology and Molecular Biology Reviews* 68(2): 173–186.

References to Chapter 2

Alemseged, Z., F. Spoor, W.H. Kimbel, R. Bobe, D. Geraads, D. Reed and J.G. Wynn. (2006). A Juvenile Early Hominin Skeleton from Dikika, Ethiopia. *Nature* 443: 296–301.

Asfaw, B., T.D. White, O. Lovejoy, B. Latimer, S. Simpson and G. Suwa. (1999). *Australopithecus garhi*: a New Species of Early hominin from Ethiopia. *Science* 284: 629–635.

Asfaw, B., W.H. Gilbert, Y. Beyene, W.K. Hart, P.R. Renne, G. WoldeGabriel, E.S. Vrba and T.D. White. (2002). Remains of Homo erectus from Bouri, Middle Awash, Ethiopia. *Nature* 416: 317–320.

Biver, N.,D. Brocklee-Morvan, J. Crovisier, et al. (2000). Spectra of Comet C/1999 (Lee). *Astronomical Journal* 120: 1554–1570.

Blackmond, D.G. (2010). The Origin of Biological Chirality. *Cold Springs Harbor Perspectives in Biology* 2010;2:a002147. Available online at http://cshperspectives.cshlp.org/content/2/5/a002147.full.pdf.

Blumenschine, R.J., F.T. Masao, J.C. Tactikos and J.I. Ebert. (2008). Effects of Distance from Stone Source on Landscape-Level Variation in Oldowan Artifact Assemblages in the Palaeo-Olduvai Basin, Tanzania. *Journal of Archaeological Science* 35: 76–86.

Brain, C.K. (1981). *The Hunters or the Hunted*. Chicago: University of Chicago Press.

Brantingham, P. J. (1998). Hominin–Carnivore Coevolution and Invasion of the Predatory Guild. *Journal of Anthropological Archaeology* 17: 327–353

Brunet, M. et al. (2002). A New Hominin from the Upper Miocene of Chad, Central Africa. *Nature* 418: 145–151.

Clarke, R.J. and Kathleen Kuman. (2000). *The Sterkfontein Palaeontological and Archaeological Site*. Johannesburg: University of Witwaterstrand Press.

Committee on the Earth System Context for Hominin Evolution. (2010). Understanding Climate's Influence on Human Evolution. Washington, DC: National Academies Press.

d'Errico, F. and L.R. Blackwell. (2003). Possible Evidence of Bone Tool Shaping by Swartkrans Early Hominins. *Journal of Archaeological Science* 30: 1559–1576.

DeHeinzlin, J., J.D. Clark, T. White, W. Hart, P. Renne, G. WoldeGabriel, Y. Beyene and E. Vrba. (1999). Environment and Behavior of 2.5-million Year Old Bouri hominins. *Science* 284: 625–629.

Donald, M. (1991). *Origins of the Modern Mind: Three Stages in the Evolution of Culture and Cognition*. Cambridge, MA: Harvard University Press.

Donald, M. (1993). Precis of Origins of the Modern Mind: Three Stages in the Evolution of Culture and Cognition. *Behavioral & Brain Sciences* 16(4): 737–791.

Donald, M. (2001). *A Mind So Rare: The Evolution of Human Consciousness*. New York: Norton.

Dunbar, R.I.M. (1992). Neocortex Size as a Constraint on Group Size in Primates. *Journal of Human Evolution* 20: 469–493.

Eales, S. (2009). *Planets and Planetary Systems*. London: Wiley–Blackwell.

Fedonkin, M.A. and B.M. Waggoner. (1997). The Late Precambrian Fossil *Limberella* is a Mullosc-Like Bilaterian Organism. *Nature* 338: 868–871.

Gamble, C. (1999). The Neanderthal in All of Us. Inaugural Lecture at the University of Southampton. Text online at the Univeristy of Southampton's Centre for the Archaeology of Human Origins: http://www.arch.soton.ac.uk/prospectus/caho/.

Gibbons, A. (2009). Of Tools and Tubers. *Science* 324(5927): 588–589.

Grine, F. (1981). Trophic Differences Between 'Gracile' and 'Robust' Australopithecines: A Scanning Electron Microscope Study of Occlusal Events. *South African Journal of Science* 77: 203–330.

Grine, F.E., W.L. Jungers and J. Schultz. (1996). Phenetic Affinities Among Early Homo Crania from East and South Africa. *Journal of Human Evolution* 30(3): 189–225.

Guthrie, R.D. (2007). Haak en Steek – The Tool That Allowed Hominins to Colonize the African Savanna and to Flourish There. pp. 133–163 in Roebrecks, W. (ed). 2007. *Guts and Brains: An Integrative Approach to the Hominin Record*. Leiden: Amsterdam University Press.

Guttman, B., A. Griffiths, D. Suzuki, and T. Cullis. *Genetics: A Beginner's Guide*. Oneworld: Oxford.

Hamilton, W.D., R. Axelrod and R. Tanese. (1990). Sexual Reproduction As An Adaptation to Resist Parasites (A Review). *Proceedings of the National Academy of the Sciences* 87: 3566–3573.

Haeusler, M. and H. McHenry. (2004). Body Proportions of *Homo habilis* Reviewed. *Journal of Human Evolution* 46: 433–465.

Holloway, R.L. (1983). Human Brain Evolution: A Search for Units, Models and Synthesis. *Canadian Journal of Anthropology* 3(2): 215–230.

Horneck, G., C. Mileikowski, H.J. Melosh, J.W. Wilson, F.A. Cucinotta and B. Gladman. (2002). Viable Transfer of Microorganisms in the Solar System and Beyond. pp. 57–76 in Hornbeck, G. and C. Baumstarck-Khan (eds). *Astrobiology: The Quest for the Conditions of Life*. New York: Springer.

Johanson, D.C. and M. A. Edey. (1981). *Lucy: The Beginnings of Humankind*. New York: Simon and Schuster.

Kaiser, D. (2001). Building a Multicellular Organism. *Annual Review of Genetics* 35: 103–123.

Keeley L.H. and N. Toth. (1981). Microwear Polishes on Early Stone Tools from Koobi-Fora, Kenya. *Nature* 293: 464–465.

Kimbel, W.H., R.C. Walter, D.C. Johanson, K.E. Reed and J.L. Arnonson. (1996). Late Pliocene *Homo* and Oldowan Tools from the Hadar Formation (Kada Hadar Member), Ethiopia. *Journal of Human Evolution* 31(6): 549–561.

Klein, R.G. (1989). *The Human Career; Human Biological and Cultural Origins*. Chicago: University of Chicago Press.

Kuman, K. and R.J. Clarke. (2000). Stratigraphy, Artefact Industries and hominin Associations for Sterkfontein, Member 5. *Journal of Human Evolution* 38(6): 827–847.

Leakey, L.S.B., P.V. Tobias and J.R. Napier. (1964). A New Species of the Genus *Homo* from Olduvai Gorge. *Nature* 202: 7–9.

Leakey, M.D. (1971). Olduvai Gorge: Excavations in Beds I & II, 1960–1963. Cambridge: Cambridge University Press.

Leakey, M.D. and R.L. Hay. (1979). Pliocene Footprints in the Laetolil Beds at Laetoli, Tanzania. *Nature* 262: 460–465.

Leakey, M.G., C.S. Feibel, I. McDougall and A. Walker. (1995). New Four-Million Year-Old Species from Kanapoi and Allia Bay, Kenya. *Nature* 376: 565–571.

Lenton, T. and A. Watson. (2011). *Revolutions that Made the Earth*. Oxford: Oxford University Press.

Lovejoy, O.C. (2009). Reexamining Human Origins in Light of *Ardipithecus ramidus*. *Science* 326: 74e1–74e8.

Margulis, L., and D. Sagan. (2002). *Acquiring Genomes: A Theory of the Origin of Species*. New York: Basic Books.

Marshall, C. R. (2006). Explaining the Cambrian 'Explosion' of Animals. *Annual Review of Earth and Planetary Sciences* 34: 355–84.

McHenry, H. (1991). Femoral Lengths and Stature in Plio-Pleistocene hominins. *American Journal of Physical Anthroplogy* 85: 149–158.

McHenry, H., and K. Coffing. (2000). Australopithecus to Homo: Transformations in Body and Mind. *Annual Review of Anthropology* 129–146.

Minkoff, E. C. (1983). Evolutionary Biology. Reading, MA: Addison–Wesley.

Mithen, S. (1996). *The Prehistory of the Mind: The Cognitive Origins of Art, Religion and Science*. London: Thames and Hudson.

Niklas, K.J. (1994). Morphological Evolution through Complex Domains of Fitness. *Proceedings of the National Academy of Sciences of the United States of America* 91: 6772–6779.

Pääbo, S. (2004). *Ancient DNA*. pp. 68–87 in Krude, T. (ed). DNA: Changing Science and Society. Cambridge: Cambridge University Press. .

Pilbeam, D. (1998). Afterword. In Akazawa, T., Aoki, K., and Bar-Yosef, O. (eds) *Neanderthals and Modern Humans in Western Asia* pp. 523–527. New York: Plenum Press.

Raup, D.M. (1991). *Extinction: Bad Genes or Bad Luck?* New York: Norton.

Potts, R. (1988). *Early Hominin Activities at Olduvai*. New York: Aldine.

Rodman, P.S. and H.M. McHenry. (1980). Bioenergetics and the Origin of Hominin Bipedalism. *American Journal of Physical Anthropology* 52(1): 103–106.

Ruiz-Mizaro, K., J. Pereto and A. Moreno. (2004). A Universal Definition of Life: Autonomy and Open-Ended Evolution. *Origins of Life and Evolution in the Biosphere* 34: 323–346.

Sagan, C. (1980). *Cosmos*. New York: Random House.

Santos, M. (1998). Origin of Chromosomes in Response to Mutation Pressure. *The American Naturalist* 152(5): 751–756.

Schenk, F., O. Kullmer and T. Bromage. (2007). The Earliest Putative Homo Fossils. *Handbook of Palaeoanthropology* 3: 1611–1631.

Semaw, S. (2000). The World's Oldest Stone Artifacts from Gona, Ethiopia: Their Implications for Understanding Stone Technology and Patterns of Human

Evolution Between 2.6–1.5 Million Years Ago. *Journal of Archaeological Science* 27: 1197–1214.

Senut, B., M. Pickford, D. Gommery, P. Mein, K. Cheboi and Y. Coppens. (2001). First hominin from the Miocene (Lukeino Formation,Kenya). *Comptes Rendus* 332: 137–144.

Secker, J., J.R. Lepcock and P.S. Wesson. (1996). Astrophysical and Biological Constraints on Radiopanspermia. *Astrophysics Online.* Available at http://arxiv.org/abs/astro-ph/9607139.

Shumaker, R.W., K. R. Walkup and B.J. Beck. (2011). *Animal Tool Behavior: The Use and Manufacture of Tools by Animals.* Baltimore: Johns Hopkins University Press.

Smith, J.M. and E. Szathmary. (1995). *The Major Transitions in Evolution.* Oxford: Oxford University Press.

Sterelny, K. (2011). Evolvability Reconsidered. pp. 83–100 in Calcott, B. and K. Sterelny (eds). *The Major Transitions in Evolution Revisited.* Vienna Series in Theoretical Biology. Cambridge, MA: MIT Press.

Stout, D., S. Semaw, M.J. Rogers and D. Cauche. (2010). Technological Variation in the earliest Oldowan from Gona, Afar, Ethiopia. *Journal of Human Evolution* 58: 474–491.

Strasdeit, H. (2010). Chemical Evolution and Early Earth's and Mars' Environmental Conditions. *Palaeodiversity* 3 (Supplement): 107–116.

Sussman, R. (2008). Brief communication: Evidence bearing on the status of *Homo habilis* at Olduvai Gorge. *American Journal of Physical Anthropology* 137(3): 356–361.

Szathmary, E. and C. Fernando. (2011). Concluding Remarks. pp. 301–310 in Calcott, B. and K. Sterelny (eds). 2011. *The Major Transitions in Evolution Revisited.* Vienna Series in Theoretical Biology. Cambridge, MA: MIT Press.

Toth, N. (1997). The Artefact Assemblages in the Light of Experimental Studies. pp. 363–401 in Isaac, G.L. and B. Isaac (eds) *Koobi Fora Research Project Vol V: Plio-Pleistocene Archaeology.* Oxford: Clarendon Press.

Ungar, P.S., F.E. Grine and M.F. Teaord. (2008). Dental Microwear and Diet of the Plio-Pleistocene Hominin Paranthropus boisei. *PLOS Online* 3(4): e2044.

Vaughan, C.L. (2003). Theories of Bipedal Walking: an Odyssey. *Journal of Biomechanics* 36: 513–523.

Walker, A.C., M.R. Zimmerman and R.E.F. Leakey. (1982). A Possible Case of Hypervitaminosis A in *Homo erectus. Nature* 296: 248–250.

Walker, A.C., R.E.F. Leakey, J.M. Harris and F.H. Brown. (1986). 2.5-Myr. *Australopithecus boisei* From West of Lake Turkanan, Kenya. *Nature* 322: 517–522.

Watson, J.D. with J. Berry. (2003). *DNA: The Secret of Life.* New York: Alfred A. Knopf.

West, S. A., S. P. Diggle, A. Buckling, A. Gardner, and A. S. Griffin. (2007). The Social Lives of Microbes. *Annual Review of Ecology, Evolution, and Systematics* 38: 53–77.

White, T.D., B. Asfaw, Y. Beyene, Y. Hailie-Selassie, C.O. Lovejoy, G. Suwa, and

F.C. Howell. (2009). *Ardipithecus ramidus* and the Palaeobiology of the Early Hominins. *Science* 326: 75–86.

Worden, A. and F. French. (2011). *Falling to Earth: An Apollo 15 Astronaut's Journey*. Washington, DC: Smithsonian Books.

References for Chapter 3

Ambrose, S.H. (1998). Late Pleistocene Human Population Bottlenecks, Volcanic Winter, and Differentiation of Modern Humans. *Journal of Human Evolution* 34: 623–651.

Bajema, C.J. (1971). *Natural Selection in Human Populations: The Measurement of Ongoing Genetic Evolution in Contemporary Societies*. New York: John Wiley & Sons.

Brodie, E.D. (2005). Caution: Niche Construction Ahead. *Evolution* 59(1): 249–251.

Childe, V.G. (1950). The Urban Revolution. *Town Planning Review* 21: 3–17.

Donald, M. (1993). *On the Origins of the Modern Mind: Three Stages in the Evolution of Culture and Cognition*. Cambridge, MA: Harvard University Press.

Donald, M. (2004). *The Definition of Human Nature*. pp. 34-58 in Rees, D. and S. Rose (eds). The New Brain Sciences: Perils and Prospects. Cambridge, Cambridge University Press.

Dunbar, R.I.M. (1993). Coevolution of Neocortical Size, Group Size and Language in Humans. *Behavioral and Brain Sciences* 16(4): 681–735.

Gabora, L. (2006). Conceptual Closure: How Memories are Woven into an Interconnected Worldview. *Annals of the New York Academy of Sciences*. Online at DOI:10.1111/j.1749-6632.2000.tb06264.x.

Koestler, A. (1964). *The Act of Creation*. London: Penguin.

Laland, K.N. and G.R. Brown. (2006). Niche Construction, Human Behavior, and the Adaptive Lag Hypothesis. *Evolutionary Anthropology* 15: 95–104.

Mithen, S. (1999). *The Prehistory of the Mind: The Cognitive Origins of Art, Religion and Science*. London: Thames and Hudson.

Moran, E.F. (1979). *Human Adaptability: An Introduction to Ecological Anthropology*. N. Scituate, MA: Duxbury Press.

Odling-Smee, F.J., K.N. Laland and M. Feldman. (2003). *Niche Construction: the Neglected Process in Evolution*. Princeton: Princeton University Press.

Palmer, S.K., L.G.Moore and D. Young. (1999). Altered Blood Pressure Course During Normal Pregnancy and Increased Preeclampsia at High Altitude (3100 meters) in Colorado. *American Journal of Obstetrics and Gynecology* 189: 1161–1168.

Pilbeam, D. (1998). *Afterword*. In Akazawa, T., Aoki, K., and Bar-Yosef, O. eds. 1998. *Neanderthals and Modern Humans in Western Asia*: 523–527. New York: Plenum Press.

Roberts, D.F. (1973). Climate and Human Variability. *Addison-Wesley Readings in Anthropology* 34. New York: Addison-Wesley.

Smith, B.D. (2007). Niche Construction and the Behavioral Context of Plant and Animal Domestication. *Evolutionary Anthropology* 16: 188–199.

References to Chapter 4

Adamsky, V. and Smirnov, Y. (1994). Moscow's Biggest Bomb: the 50-Megaton Test of October 1961. *Cold War International History Project Bulletin*, Issue 4, pp. 3, 19–21.

Alvarez, W. (1997). *T Rex and the Crater of Doom*. Princeton: Princeton University Press.

Asimov, I. (1979). *A Choice of Catastrophes*. New York: Ballantine.

Becker, L. *et al.* (2004). Bedout: A Possible End-Permian Impact Crater Offshore of Northwestern Australia. *Science* 304: 1469–1476.

Boorstin, D. (1985) *The Discoverers: A History of Man's Search to Know his World and Mimself*. New York: Vintage Books.

Bostrom, N. (2007). *The Future of Humanity*. Future of Humanity Institute, Oxford University.

Coe, M. (1993). *The Maya*. 5th Edition. New York: Thames and Hudson.

DeGroot, G. (2005). *The Bomb: A Life*. Cambridge, MA: Harvard University Press.

Diamond, J. (2005). *Collapse: How Societies Choose to Fail or Succeed*. New York: Viking.

Ganshof, F-L. (translated by P. Grierson). (1964). *Feudalism*. 3rd edition. London: Longmans.

Grabois, A. (1980). *The Illustrated Encyclopedia of Medieval Civilization*. Mayfield Books: New York.

Grieve, R. A. F. (1994). The Economic Potential of Terrestrial Impact Craters. *International Geology Review* 36 (2): see DOI: 10.1080/00206819409465452.

Jackson, J.C.B. et al. (2001). Historical Overfishing and the Recent Collapse of Coastal Ecosystems. *Science* 293: 629–637.

James Martin 21st Century School. (2008). *Policy Foresight and Global Catastrophic Risks*. London.

Kaiho, K. *et al.* (2001). End-Permian Catastrophe by a Bolide Impact: Evidence of a Gigantic Release of Sulfur from the Mantle. *Geology* 29(9): 815–818.

Levathes, L. (1994). *When China Ruled the Seas: The Treasure Fleet of the Dragon Throne 1405–1433*. New York: Oxford University Press.

Levin, H. (2010). *The Earth through Time* (9th edn). Hoboken, NJ: Wiley and Sons.

Lewis, J. (1997). *Mining the Sky: Untold Riches from the Asteroids, Comets and Planets*. New York: Helix Books.

Morrison, D., Chapman, C. R., Steel, D., and Binzel R. P. (2004). Impacts and the Public: Communicating the Nature of the Impact Hazard. In Belton,M.J.S., T.H. Morgan, N.H. Samarasinha and D.K. Yeomans (eds). 2004. *Mitigation of Hazardous Comets and Asteroids*. Cambridge: Cambridge University Press.

Needham, J. (1991). *Science and Civilization in China*. Volume 4: Physics and

Physical Technology (II): Mechanical Engineering. Cambridge: Cambridge University Press.

O'Keefe, B. (1983). *Nuclear Hostages*. Boston: Houghton Mifflin.

Ramenofsky, A.F. (1987). *Vectors of Death: The Archaeology of European Contact*. Albuquerque, NM: University of New Mexico Press.

Rees, M. (2003). *Our Final Hour: A Scientists Warning. How Terror, Error and Environmental Disaster Threaten Humankind's Future in this Century and Beyond*. New York: Basic Books.

Regis, E. (2000). *The Biology of Doom: The History of America's Secret Germ Warfare Program*. New York: Holt.

Science Daily (2011, Jan. 25). See www.sciencedaily.com/releases/2011/01/110123131014.htm

Shapiro, R. (1999). *Planetary Dreams: The Quest to Discover Life Beyond Earth*. New York: Wiley.

Shin, S., & Shi, G. (2002). Paleobiogeographical Extinction Patterns of Permian Brachiopods in the Asian-Western Pacific Region. *Paleobiology* 28: 449–463.

Tainter, J. (1990). *The Collapse of Complex Societies*. Cambridge: Cambridge University Press.

Wignall, P. & Twitchett, R. (2002). Permian–Triassic sedimentology of Jameson Land, East Greenland: Incised submarine channels in an anoxic basin. *Journal of the Geological Society* 159(6): 691–703.

Wilson, S. (1992). The Emperor's Giraffe. *Natural History* 101(12): 22-26.

Zubrin, R. (1999). *Entering Space: Creating a Spacefaring Civilization*. New York: Tarcher/Penguin New York.

References to Chapter 5

Anderson, G. and B. Roskrow. (1994). *The Channel Tunnel Story*. London: E. & F.N. Spoon.

Bell. J.F. (2005). *The Dream Palace of the Space Cadets*. Op-ed article at *Spacedaily.com,* available at http://www.spacedaily.com/news/oped-05zzb.html.

Bennett, J., S. Shostak and B. Jakosky. (2003). *Life in the Universe*. San Francisco: Addison Wesley.

Bradbury, R. (1977). The Life Force Speaks: We Move to Answer. pp. 9–10 in Heppenheimer, T.A. 1977. *Colonies in Space*. Harrisburg, PA: Stackpole Books.

Bradbury, R. (1997). *The Martian Chronicles*. New York: Harper Collins.

Cabbage, M. and W. Harwood. (2004). *Comm Check: The Final Flight of Shuttle Columbia*. New York: Free Press.

Clarke, A.C. (1999). We'll Never Conquer Space (originally published in 1960). pp. 204–210 in Clarke, A.C. 1999. *Greetings, Carbon-Based Bipeds! Essays 1939–1998*. New York: St. Martin's Press.

Daniels, N. (1986). Consent to Risk in Space. pp. 277–290 in Hargrove, E.C. (ed). 1986. *Beyond Spaceship Earth: Environmental Ethics in the Solar System*. San Francisco: Sierra Club Books.

DeGroot, G.J. (2006). *Dark Side of the Moon: The Magnificent Madness of the American Lunar Quest*. New York: New York University Press.

DeWaal, A. (2009). *Famine Crimes: Politics & the Disaster Relief Industry in Africa*. Bloomington, IN: Indiana University Press.

Eales, S. (2009). *Planets and Planetary Systems*. Chichester: Wiley-Blackwell.

Eckart, P. (1994). *Spaceflight Life Support and Biospherics*. Dordrecht: Kluwer Academic.

Galton Institute. (2005). *Third World Aid and Contraception: Either or Both?* http://www.galtoninstitute.org.uk/Newsletters/GINL9105/Third_World_Aid.htm

General Accounting Office. (2011). *Defense Acquisitions: Assessments of Selected Weapon Programs*. Available at http://www.gao.gov/products/GAO-11-233SP.

Gert, B. (2004). *Common Morality: Deciding What to Do*. New York: Oxford University Press.

Guilane, J. and J. Zammit. (2005). *The Origins of War: Violence in Prehistory*. Oxford: Wiley-Blackwell.

Hodges, W. (1985). The Division of Labor and Interstellar Migration: A Response to 'Demographic Contours'. pp. 134–151 in Finney, B.R. and E.M. Jones (eds). *Interstellar Migration and the Human Experience*. Berkeley, CA: University of California Press.

Japan Aerospace Exploration Agency. (2003). *Space Medicine*. Website at http://iss.jaxa.jp/med/index_e.html.

Keely, L. (1996). *War Before Civilization: The Myth of the Peaceful Savage*. Oxford: Oxford University Press.

Kelly, R.C. (2000). *Warless Societies and the Origin of War*. Chicago: University of Chicago Press.

Knapp, B. (1989). *Machine, Metaphor and the Writer: A Jungian View*. Pennsylvania, PA: Pennsylvania State University Library.

Lininger, J. (2000). *Off the Planet: Surviving Five Perilous Months Aboard the Space Station Mir*. New York: McGraw-Hill.

Mailer, N. (1970). *Of a Fire on the Moon*. Boston: Little, Brown and Co.

Mallan, L. (1955). *Men, Rockets and Space Rats*. New York: Messner.

McCurdy, H.E. (1997). *Space and the American Imagination*. Washington, DC: Smithsonian Institution Press.

McGovern, T.H., G. Bigelow, T. Amorosi and D. Russell. (1988). Northern Islands, Human Error, and Environmental Degradation: A View of Social and Ecological Change in the Medieval North Atlantic. *Human Ecology* 16(3): 225–270.

Miller, B. (2011). *One Small Step Back for Man. London Times* 'Eureka' supplement, March 2011.

National Research Council: Ad Hoc Committee on the Solar System Radiation Environment and NASA's Vision for Space Exploration. (2008). *Space Radiation Hazards and the Vision for Space Exploration*. Washington, DC: National Academies Press.

Sirgany, A. (2009). *The War on Waste*. Available online at http://www.cbsnews.com/stories/2002/01/29/eveningnews/main325985.shtml

Task Group on Life Sciences: Space Science Board. (1988). *Space Science in the Twenty-First Century: Imperatives for the Decades 1995 to 2015.* Washington, DC: National Academy Press.

Tribble, A.C. (2003). *The Space Environment: Implications for Spacecraft Design.* Princeton: Princeton University Press.

United Nations. (2004). *World Population to 2300.* Available online at: www.un.org/esa/population/publications/ ... /WorldPop2300final.pdf.

United Nations. (2005). *United Nations 2005 World Summit Outcomes.* New York: United Nations.

United Nations. (2010). *United Nations Millennium Development Goals Fact Sheet.* Available online at http://www.un.org/millenniumgoals/pdf/MDG_FS_2_EN.pdf.

Vaughan, D. (1996). *The Challenger Launch Decision: Risky Technology, Culture, and Deviance at NASA.* Chicago: University of Chicago Press.

Varley, P. (1994). History of the Geological Investigations for the Channel Tunnel. pp. 5–18 in Harris, C.S., M.B. Hart, P.M. Varley and C.D. Warren. *Engineering Geology of the Channel Tunnel.* London: Thomas Telford.

White, F. (1998). *The Overview Effect: Space Exploration and Human Evolution.* Reston, VA: American Institute for Aeronautics and Astronautics.

Zubrin, R. (1997). *The Case for Mars: The Plan to Settle the Red Planet and Why We Must.* New York: The Free Press.

References to Chapter 6

Alexander, P.F. (1916). *The Earliest Voyages Round the World (1519–1617).* Cambridge: Cambridge University Press.

Ambrose, W. and R.C. Green. (1972). First Millennium BC Transport of Obsidian from New Britain to the Solomon Islands. *Nature* 237: 31.

Ammarell, G. (1999). Bugis Navigation. *Yale Southeast Asia Studies Monograph* 48. New Haven, CT: Yale University Southeast Asia Studies.

Beaglehole, J.C. (1955). *The Journals of Captain James Cook on His Voyages of Discovery: Vol 1. The Voyage of the* Endeavour *1768–1771.* Cambridge: Cambridge University Press for the Hakluyt Society.

Bellwood, P. (1978). *The Polynesians: Prehistory of an Island People.* London: Thames and Hudson.

Bellwood, P. (1993). The Origins of Pacific Peoples. pp. 2–14 in Quanchi, M. and R. Adams (eds). *Culture Contact in the Pacific.* Cambridge: Cambridge University Press.

Biggs, B.G. and M. V. Biggs. (1975). Na Ciri Kalia. *Working Papers in Anthropology, Archaeology, Linguistics and Maori Studies* #42. Auckland: University of Auckland.

Buck, P.H. (1938). *Vikings of the Pacific.* Philadelphia: Lippincourt.

Buck, P.H. (1950). Material Culgture of Kapingmarangi. *Bernice P. Bishop Museum Bulletin* 200. Honolulu: Bishop Museum.

Cook, J. (1777). *A Voyage Towards the South Pole and Round the World . . . . in the years 1772–1775*. London: Strahan and Cadell. Vols I, II.

Corney, B.G. (1919). *The Quest and Occupation of Tahiti by Emissaries of Spain During the Years 1772–1776*. Vols 1–3. London: The Hakluyt Society.

Dodd, E. (1972). *Polynesian Seafaring: A Disquisition on Prehistoric Celestial Navigation and the Nature of Seagoing Double Canoes with Illustrations Reproducing Original Field Sketches, Wash Drawings, or Prints by Artists on Early Voyages of Exploration and Occasional Written Reports from On-the-Scene Observers*. Lymington, Hampshire: Nautical Publishing Company.

d'Urville, J.S-C. D. (1832). Sur le iles du Grand Ocean. *Bulletin de la Societe de Geographie* 17. (Translated by B. Douglas).

Evans, B.M. (2008). Simulating Polynesian Double-Hulled Canoe Voyaging. pp. 143–153 in Di Piazza, A. and E. Pearthree (eds) *Canoes of the Grand Ocean*. British Archaeological Reports International Series 1802. Oxford: Archaeopress.

Feinberg, R. (1988). *Polynesian Seafaring and Navigation: Ocean Travel in Anutan Culture and Society*. Kent, OH: Kent State University Press.

Finney, B. (1977). Voyaging Canoes and the Settlement of the Pacific. *Science* 196: 1277–1285.

Finney, B. (1985). *Voyagers Into Ocean Space*. pp. 164–179 in Finney, B. and E.M. Jones (eds). 1985. Interstellar Migration and the Human Experience. Berkeley, CA: University of California Press.

Finney, B. (1991). Myth, Experiment, and the Re-Invention of Polynesian Voyaging. *American Anthropologist* 87: 9–26.

Finney, B. (1994). *Voyage of Rediscovery: A Cultural Odyssey Through Polynesia*. Berkeley, CA: University of California Press.

Gamble, C. (1996). *Timewalkers: The Prehistory of Global Colonization*. New York: Harvard University Press.

Golson, J. (1977). No Room at the Top: Agricultural Intensification in the New Guinea Highlands. pp. 601–638 in Allen, J. et al. (eds). *Sunda and Sahul*. London: Academic Press.

Groves, C. (1981). *Ancestors for the Pigs: Taxonomy and Phylogeny of the Genus* Sus. Technical Bulletin 3, Research School of Pacific Studies. Canberra: Australia National Museum.

Hawkes, T. (1977). *Semiotics and Structuralism*. London: Methuen.

Horridge, A. (2008). Origins and Relationships of Pacific Canoes and Rigs. pp. 105 in Di Piazza, A. and E. Pearthree (eds) *Canoes of the Grand Ocean*. British Archaeological Reports International Series 1802. Oxford: Archaeopress.

Huffman, K.W. (1996). *Su tuh netan'monbwe:* we write on the ground: Sand Drawings and their Associations in Northern Vanuatu. pp. 247–253 in Bonnemaison, J., C. Kaufmann, K. Huffman and D. Tryon (eds). *Arts of Vanuatu*. Bathurst, UK: Crawford House.

Hutchins, E. (1995). *Cognition in the Wild*. Cambridge, MA: MIT Press.

Irwin, G. (1992). *The Prehistoric Exploration and Colonization of the Pacific*. Cambridge: Cambridge University Press.

Irwin, G. (1998). The Colonization of the Pacific Plate: Chronological, Navigational and Social Issues. *Journal of the Polynesian Society* 107(2): 111–143.

Irwin, G. and S. Blacker. (199)0. Voyaging by Canoe and Computer: Experiments in the Settlement of the Pacific. *Antiquity* 64: 34–50.

Kayser, M., S. Brauner, G. Weiss, P.A. Underhill, L. Roewer, W. Schiefenho-umlaut-vel and M.F. Hammer. (2000). Melanesian Origin of Polynesian Y Chromosomes. *Current Biology* 10: 1237–1246.

Kirch, P.V. (1997). *The Lapita Peoples: Ancestors of the Oceanic World*. Oxford: Blackwell.

Levison, M., R.G. Ward and J.W. Webb. (1973). *The Settlement of Polynesia: A Computer Simulation*. Minneapolis, MN: University of Minnesota Press.

Lewis, D. (1972). *We, the Navigators: the Ancient Art of Landfinding in the Pacific*. Canberra: Australian National University Press.

Merriwether, D.A., J.S. Friedlander, J. Mediavilla, C. Mgone, F. Gentz and R.E. Ferrell. (1999). Mitochondrial DNA Variation is an Indicator of Austronesian Influence in Island Melanesia. *American Journal of Physical Anthropology* 110: 243–270.

Nunn, P. (1993). Facts, Fallacies, and the Future in the Island Pacific. pp. 112–115 in Waddell, E., V, Naidu and E. Hau'ofa. *A New Oceania: Rediscovering Our Sea of Islands*. Suva, FIji: School of Social and Economic Development, University of the South Pacific.

Salmond, A. (2008). Voyaging Exchanges: Tahitian Pilots and European Navigators. pp. 23–46 in Di Piazza, A. and E. Pearthree (eds) *Canoes of the Grand Ocean*. British Archaeological Reports International Series 1802. Oxford: Archaeopress.

Sinoto, Y.H. and P.C. McCoy. (1975). Report on the Preliminary Excavation of an Early Habitation Site on Huahine, Society Islands. *Journal de la Societe des Oceanistes* 31: 143–186.

Smith, C.M., J.F. Haslett,G. Baker and I. Lopez. (2006). On the Vessel Sailed By Bartholome Ruiz in 1526: Characterization and Significance for the Pre-columbian Maritime Activities of Coastal Ecuadoreans. *Terrae Incogniate* 33: 55–88.

Vacella, R. (2008). Construction of Dugout and Sewn Plank Canoes on Raivavae, Austral Islands. pp. 107–120 in Di Piazza, A. and E. Pearthree (eds) *Canoes of the Grand Ocean*. British Archaeological Reports International Series 1802. Oxford: Archaeopress.

Webb, S.G. (2006). *The First Boat People*. Cambridge: Cambridge University Press.

References to Chapter 7

Alland, A. (1970). *Adaptation in Cultural Evolution: An Approach to Medical Anthropology*. New York: Columbia University Press.

Alterhaug, B. (2009). Creativity and Improvisation as Phenomena and Acting Potential in Different Contexts. pp. 161–170 in Lang, M.A. and A.O. Brubakk (eds). *The Future of Diving: 100 Years of Haldane and Beyond*. Washington, DC: Smithsonian Institution Press.

Altschuller, G. (1998). *40 Principles: TRIZ Keys to Technical Innovation*. Translated by L. Shulyak. Worcester, MA: Technical Innovation Center, Inc.

Arthur, W.B. (2006). *The Nature of Technology: What it Is and How it Evolves*. London: Allen Lane/Penguin.

Benyus, J.M. (1997). *Biomimicry*. New York: William Morrow.

Collins, H. and T. Pinch. (1998). *The Golem at Large: What You Should Know About Technology*. Cambridge: Cambridge University Press.

Dertouzos, M.I. (2002). Human-Centered Systems. pp. 181–191 in Denning, P.J. (ed) *The Invisible Future: the Seamless Integration of Technology Into Everyday Life*. New York: McGraw-Hill.

Dyke, C. (1988). Cities as Dissipative Structures. pp. 355–367 in Weber, B., D.J. Depew and J.D. Smith (eds). *Entropy, Information, and Evolution: New Perspectives in Physical and Biological Evolution*. Cambridge, MA: MIT Press.

Eiben, A.E. and J.E. Smith. (2008). *Introduction to Evolutionary Computing*. New York: Springer.

Fuad-Luke, A. (2010). Adjusting Our Metabolism: Slowness and Nourishing Rituals of Delay in Anticipation of a Post-Consumer Age. pp. 133–155 in Cooper, T. (ed). *Longer Lasting Products: Alternatives to the Throwaway Society*. Surrey: Gower.

Grant, V. (1963). *The Origin of Adaptations*. New York: Columbia University Press.

Heckler, R.S. (2002). Somatics in Cyberspace. pp. 277–294 in Denning, P.J. (ed). *The Invisible Future: The Seamless Integration of Technology into Everyday Life*. New York: McGraw-Hill.

Hodges, W.A. (1985). The Division of Labor and Interstellar Migration: A Response to 'Demographic Contours'. pp. 134–151 in Finney, B.R. And E.M. Jones (eds). *Interstellar Migration and the Human Experience*. Berkeley, CA: University of California Press.

Janis, I. (1982). *Groupthink: Psychological Studies of Policy Decisions and Fiascoes*. 2nd ed. Boston, MA: Houghton Mifflin.

Kehoe, A. (1985). Modern Antievolutionism: The Scientific Creationists. pp. 156–185 in Godfrey, L.R. (ed). What Darwin Began: Modern Darwinian and Non-Darwinian Perspectives on Evolution. Boston, MA: Allyn & Bacon.

Kuhn, T. (1962). *The Structure of Scientific Revolutions*. Chicago: University of Chicago Press.

Levinson, P. (2003). *Realspace: the Fate of Physical Presence in the Digital Age, On and Off Planet*. London: Routledge.

Losos, J.B., and D.L. Mahler. (2010). Adaptive Radiation: The Interaction of Ecological Opportunity, Adaptation, and Speciation. pp. 381–420 in Bell, M.A., D.J. Futuyma, W.F. Eanes, and J.S. Levinton (eds). *Evolution Since Darwin: The First 150 Years*. Sunderland, MA: Sinauer.

Miller, B. (2011). *One Small Step Back for Man*. London Times, 'Eureka' supplement, P. 58.

Monbiot, G. (2003). *With Eyes Wide Shut*. Online at http://www.guardian.co.uk/environment/2003/aug/12/comment.columnists.

Murdock, G.P. (1945). The Common Denominator of Cultures. pp. 123–125 in Linton, R. (ed). *The Science of Man in the World Crisis*. New York: Columbia University Press.

National Academy of Engineering. (2010). *The Engineer of 2020: Visions of Engineering in the New Century*. Washington, DC: National Academies Press.

Nauman, R.B., A.M. Dellinger, E. Zaloshjna, B.A. Lawrence and T.R. Miller. (2010). Incidence and Total Lifetime Costs of Motor Vehicle-Related Fatal and Non-Fatal Injury by Road User Type in the United States. *Injury Prevention* 16. Online at doi:10.1136/ip.2010.029215.116.

Petroski, H. (1985). *To Engineer is Human: the Role of Failure in Successful Design*. New York: St. Martin's Press.

Prosser, C.L. (1986). *Adaptational Biology: Molecules to Organisms*. New York: Wiley.

Redman, C.L. (2001). *Human Impact on Ancient Environments*. Phoenix, AZ: University of Arizona Press.

Regis, E. Jr. (1985). The Moral Status of Multigenerational Interstellar Exploration. pp. 248–259 in Finney, B.R. And E.M. Jones (eds). *Interstellar Migration and the Human Experience*. Berkeley, CA: University of California Press.

Sawyer, K. (2008). Improvisation and Teaching. *Critical Studies in Improvisation* (3) # 2.

Simpson, G.G. (1949). *The Meaning of Evolution: A Study of the History of Life and its Significance for Man*. New Haven, CT: Yale University Press.

Sumner, F.B. (1919). Adaptation and the Problem of 'Organic Purposiveness'. *American Naturalist* 53(626): 193–217.

Vallero, D. and C. Brasier. (2008). *Sustainable Design: The Science of Sustainability and Green Engineering*. Hoboken, NJ: Wiley.

Vaughan, D. (1996). *The Challenger Launch Decision: Risky Technology, Culture, and Deviance at NASA*. Chicago: University of Chicago Press.

Vermeij, G.J. (1978). *Biogeography and Adaptation*. Cambridge, MA: Harvard University Press.

Vigor, J. (2001). *Seaworthy Offshore Sailboat: A Guide to Essential Features, Handling, and Gear*. Camden, ME: McGraw Hill.

Wallace, B. and A.M. Srb. *Adaptation*. 2nd edition. Upper Saddle River, NJ: Prentice-Hall.

West, S.A., S.P. Diggle, A. Buckling, A. Gardner and A.S. Griffin. (2007). The Social Lives of Microbes. *Annual Review of Ecology, Evolution, and Systematics* 38: 53–77.

White, M. (2005). *Life Out There: The Truth of – and Search for – Extraterrestrial Life*. New York: Little, Brown and Company.

Woese, C. (2004). A New Biology for a New Century. *Microbiology and Molecular Biology Reviews* June 2004: 173–86.

References to Chapter 8

Barlow, N. (2008). *Mars: An Introduction to its Interior, Surface and Atmosphere*. Cambridge: Cambridge University Press.

Bauman, R. (2009). *Microbiology with Diseases by Body System*. San Francisco: Pearson Benjamin.

Beech, M. (2009). *Terraforming. The Creating of Habitable Worlds*. New York: Springer.

Bibring, J-P. (2011). Water on Mars. pp. 234–246 in Gargaud, M., P. Lopez-Garcia and H. Martin (eds). *Origins and Evolution of Life; An Astrobiological Perspective*. Cambridge: Cambridge University Press.

Burton, R.F. (1872). *Zanzibar: City, Island, and Coast* (Volume 1). London: Tinsley Brothers.

Cohen, M. (2002). *Selected Precepts in Lunar Architecture*. Paris: International Astronautical Federation.

Conerly, P. (2009). *The New Moon Race: Lunar Agriculture*. Unpublished BS degree thesis, Worcester Polytechnic Institute.

Conversation with the Sidereal Messenger in *Dissertatio cum Nuncio Sidereo* (1610) printed in *Johannes Kepler Gesammelte Werke* 1937 (IV):308, ll.

Crosby, A.W. (1985). Life (With All its Problems) in Space. pp. 210–219 in Finney, B. and E.M. Jones (eds). *Interstellar Migration and the Human Experience*. Berkeley, CA: University of California Press.

Davies, P. (2010). *The Eerie Silence: Renewing our Search for Alien Intelligence*. Boston: Houghton, Mifflin.

Duke, M.B. (2006). Development of the Moon. *Reviews in Mineralogy and Geochemistry* Online at DOI: 10.2138/rmg.2006.60.6.

Folger, T. (2011). The Planet Boom. *Discover*; 32(4): 30–39.

Friedberg, E.C. (2006). Mutation as a Phenotype. Pp 39–56 in Caporale, L.H. (ed) 2006. *The Implicit Genome*. Oxford: Oxford University Press.

Levin, H. (2010). *The Earth through Time*. Hoboken, NJ: Wiley.

Gomes, R. *et al.* (2005). Origin of the Cataclysmic Late Heavy Bombardment Period of the Terrestrial Planets. *Nature* 435: 466–469.

Gilster, P. (2004). *Centauri Dreams: Imagining and Planning Interstellar Exploration*. New York: Copernicus Books.

Grinspoon, D. (2003). *Lonely Planets: The Natural Philosophy of Alien Life*. New York: Harper-Collins.

Leake, J. (2010). Don't Talk to Aliens, Warns Stephen Hawking. *The Sunday Times*. April 25. London: Times Newspapers.

Murdock, G.P. (1945). The Common Denominator of Cultures. pp. 123125 in Linton, R. (ed). *The Science of Man in the World Crisis*. New York: Columbia University Press.

NASA Mission News. (2010). *Phoenix Mars Lander: Exploring the Arctic Plain of Mars Mission News*. 25 May 2010 (http://www.nasa.gov/mission_pages).

Peters. W. (1984). The Appearance of Mars and Venus in 1610. *Journal of the History of Astronomy* 15(3): 211–214.

Poggie, J.J. And C. Gersuny. (1976). Risk and Ritual in a New England Fishing Community. pp. 351–369 in Poggie, J.J., G.H. Pelto and P.J. Pelto (eds). *The Evolution of Human Adaptations: Readings in Anthropology*. New York: Macmillan.

Rees, M. (2003). *Our Final Hour. How Terror, Error and Environmental Disaster Threaten Humankind's Future in this Century-on Earth and Beyond.* New York: Basic Books.

Roberts, C. and K. Manchester. (2005). *The Archaeology of Disease.* Stroud: The History Press.

Rosenberg E., G. Sharon and I. Zilber-Rosenberg. (2009). The Hologenome Theory of Evolution: a Fusion of neo-Darwinism and Lamarckism. *Environmental Microbiology* 11: 2959–2962.

Rosenberg, E. and Ziller-Rosenberg, I. (2011). Symbiosis and Development: the Hologenome Concept. *Birth Defects Research (Part C)* 93: 56–66.

Rosenberg, N.A., S. Mahajanm, C. Ramachandran, C. Zhao, J.K. Pritchard et al. (2005) Clines, Clusters, and the Effect of Study Design on the Inference of Human Population Structure. *PLoS Genetics* 1(6): e70. doi:10.1371/journal.pgen.0010070.

Rozwadowski, H.M. (2005). *Fathoming the Ocean.* Cambridge, MA: Belknap Press.

Sagan, C. (1980). *Cosmos.* New York: Random House.

Seager, S. (2010). The Hunt for Super Earths. *Sky and Telescope* 120(4): 30.

Sheehan, W. (1996). *The Planet Mars: A History of Observation and Discovery.* Tuscon, AZ: University of Arizona Press.

Shostak, S. (2011). A Bucketful of Worlds. *The Huffington Post.* Online at http://www.huffingtonpost.com/seth-shostak/a-bucketful-of-worlds_b_817921.html.

Spohn, T. (2003). Oceans in the Icy Galilean Satellites of Jupiter? *Icarus* 161: 456–467.

Steele, W. (1999). It's the 25th Anniversary of Earth's First (and Only) Attempt to Phone ET. *Cornell News* November 1999.

Taton, R. *et al.* (2003). *Planetary Astronomy from the Renaissance to the Rise of Astrophysics, Part A; Tycho Brahe to Newton.* Cambridge: Cambridge University Press.

Thomas, C.D. (1990). What Do Real Population Dynamics Tell Us About Minimum Viable Population Sizes? *Conservation Biology* 4(3): 324–327.

Tishkoff, S. A., F.A. Reed, A. Ranciaro, B. F. Voight, C.C. Babbitt, J.S. Silverman, K. Powell, H.M. Mortensen, J.B. Hirbo, M. Osman, M. Ibrahim, S.A. Omar, G. Lema, T.B. Nyambo, J. Ghori, S. Bumpstead, J.K. Pritchard, G.A. Wray and P. Deloukas. (2007). Convergent Adaptation of Human Lactase Persistence in Africa and Europe. *Nature Genetics* 39(1): 31–40.

Zubrin, R. (1999). *Entering Space: Creating a Spacefaring Civilization.* New York: Tarcher/Putnam.

Zubrin, R. and Wagner, R. (1997). The Case for Mars: The Plan to Settle the Red Planet and Why We Must. New York: Touchstone.

Index